Modern European Landscape Design

现代欧洲景观设计

（德）乌多·达根巴赫／编　常文心／译

辽宁科学技术出版社

Modern European Landscape Design

Those who read this book and look through it will expect to discover a difference to other landscape design books showing works from other regions. Do they really get what they expect? I think yes.

The global networking of the landscape design community is super fast today, but we still are facing the analogue realities in the countries where the works are developed for. Climate, culture, size of the countries, topographic differences, regional traditions and differences in political systems and political culture have a strong impact on the design and its results in realisation.

So we have to think about the two key words in the title of the book: Modern and European. Modern means to me: not repeating old styles or copy them, but it can be modern to reinvent them. That is what I would suggest as a simple understanding of the word. Europe is a geographic and that way also cultural defined space which I would not like to mix with political borders. The Ural River is the defined eastern geographic border of Europe and the Atlantic coast in the west. When I once stood on bridge the crossing the Ural River in the Kazak city of Atyrau defining the border between Asia and Europe I really understood how far east this border is. A quite big heterogeneous and multicultural space with many languages and cultures.

That is why I am including our colleagues from Moscow, Kiev, Tblisi, Baku, Yerevan, Istanbul as a part of the European landscape architects community as well as the one from the Baltic States, the Balkan states, the new east European states and the one of the countries of the former European mainland.

Is there a typical European style in landscape design which we could name and define? If there ever existed one, it is vanishing daily I assume. Due to fast growing databases which help to search and compare landscape designs all over the world we are getting closer together and building up a world community of landscape architects – that a kind of equalises our works as well.

The mixture of people working in European offices for landscape design as never been so wide spread. In our Berlin office we have colleagues from Italy, Georgia, Poland and Iran. Communication is English like in most mixed offices. This influences daily work and design directions.

But regarding the present status which is documented in this book we can possibly register some general directions which developed in the "big laboratory Europe".

Public spaces as bigger parks, temporary garden shows with remaining park elements in it, city squares, school yards, municipal playgrounds and other hybrid public spaces will generally show a more or less minimalistic attitude, especially in Switzerland, the Netherlands, UK,

现代欧洲景观设计

这本书的读者一定期待在本书中找到与其他地区不一样的景观设计。我相信他们一定能如愿以偿。

景观设计的全球化网络正飞速发展，但是各地的现状仍然深刻影响着景观设计。气候、文化、国家规模、地形差别、地区传统、政治体系差异以及政治文化等都对设计及设计的实施有着巨大的影响。

这样一来，我们必须对本书标题的两个关键字"现代"和"欧洲"进行思考。对我来说，"现代"意味着设计并不会重复旧风格或抄袭它们，而是会对其进行重新诠释。这也是我对它的简单理解。"欧洲"是一个地理词汇，同时也象征着一种文化，它与政治界限无关。乌拉尔河是欧洲的东部边界，而大西洋海岸则是西部边界。当我站在哈萨克斯坦阿特劳市乌拉尔河的大桥上远眺，我才真正意识到这个东部边界有多远。欧洲是一个多形态、多文化地区，融合了各种各样的语言与文化。

因此，我把来自莫斯科、基辅、第比利斯、巴库、耶烈万、伊斯坦布尔、波罗的海诸国、巴尔干半岛诸国、新东欧诸国以及传统欧洲大陆国家的同行都归入欧洲景观设计师这个行列。

有没有一种我们可以直接命名或定义的典型欧洲景观设计风格？即使有的话，我觉得它也会迅速消失。随着数据库的快速发展，我们能快速搜索和对比世界各地的景观设计。一个全球化的景观设计师团体正在集结，我们的作品也得到了快速的交流。

目前，欧洲各个景观设计事务所的设计

Germany and France.

The southern countries close to the dry and hot climate zone as well as the northern countries in the more cold and wet climate zone have developed a reductive and sometimes more architectural design style which is partly a result of the climate.

The new east European countries, the Baltic States and the Balkan states dare to design more innovative and challenging as if the new gained possibility to do so needs to be checked out. There is more curiosity to test new materials, to mix them. The style mix of the first years in the 90ies is gone meanwhile.

Generally the design for those spaces is determined through a growing participation process which includes a moderated design phase together with interested and engaged citizens. Another strong parameter are the declining financial possibilities of the communities and the cities. They need a minimalistic design which guarantees minimal maintenance costs. Landscape architecture is part of the infrastructural task, a community has to manage and so sometimes design is really poor.

Private residences show the most regional characteristic elements. Here the clients determine the direction a designer can walk. And possibly one very typical attitude for the "Old European countries" is: "Even if you are very rich, you should not show off – it arouses social envy or jealousy." That is why most designs are rich but with a modest attitude, which is most likely one of the most typical elements in European Landscape architecture. It creates a special style which is developed to a very high level in Switzerland. As Minimalism is a European child from the early 20th century most landscape architects are educated in its history and tradition and it became a regular tool of European design.

I am most likely expecting a new ornamental design style the next decades which will go back to transformed traditions from baroque and rococo period. If I am right – it will be shown in a new book ten years – if I am wrong I feel comfortable in the row of architects predicting wrong trends…

Please enjoy this book as a chance to have a look in the "Laboratory Europe"

Udo Dagenbach
April 2015, Berlin

师来自于五湖四海。在我们的柏林办公室，就有来自意大利、格鲁吉亚、波兰、伊朗等各国的同事，我们通过英语进行交流。这种融合影响了我们的日常工作和设计方向。

但是通过本书中的项目现状，我们基本可以从"大欧洲实验室"中总结出一些设计的大方向。

大型公园、临时园艺展、城市广场、校园、市政运动场以及其他混合公共空间的设计或多或少都会向极简主义倾斜，特别是在瑞士、荷兰、英国、德国、法国等地。
南欧的国家气候干燥炎热，北欧国家则较为阴冷潮湿，出于气候原因，它们的景观设计更为简约，有时也更倾向于建筑设计风格。

新东欧国家、波罗的海诸国和巴尔干半岛诸国敢于打造更具创新和挑战的设计。他们不断进行新的尝试，乐于尝试新材料并将其混合起来。但是20世纪90年代的混合风格已经一去不返了。

越来越多的相关利益者和市民开始参与到设计过程中，这在很大程度上决定了这些公共空间的设计。另一个重要的参数是社区与城市日渐衰落的财政能力。景观设计是基础建设的一部分，必须由社区进行管理，因此一些设计水平可能会让人失望。

私人住宅是最能体现区域特征的景观项目。在这里，由委托人决定设计的方向。在老牌欧洲国家有一种典型的看法："即使你很富有，也不要过分炫耀，那会引来别人的嫉妒。"这也是为什么许多设计丰富而低调的原因。丰富而低调是欧洲景观设计的一大特色，瑞士的设计将其演绎到了极致。极简主义起源于20世纪早期的欧洲，因此大多数景观建筑师都深受其影响，让极简主义成为了欧洲设计的常用手段。

我本人十分期待在未来的几十年内能出现一种新的装饰设计风格，希望它能向巴洛克和洛可可时代靠拢。如果我是对的，那么几十年后的书中就会收录这些项目；如果我错了，那么我也无所谓站在了预测失误的设计师行列。

本书将为您带来"大欧洲实验室"景观设计的风貌，敬请阅读。

乌多·达根巴赫
2015年4月，柏林

A NEW WIND IN EUROPEAN LANDSCAPE DESIGN

At the end of his life, the famous Dutch documentary maker Joris Ivens filmed his great work "Une Histoíre de vent", a Tale of the Wind. In this film Ivens is seen in his travel through the landscapes of China, trying to "film the wind as he tried all his life". This metaphor applies for the landscape-architecture as well, for is designing the landscape not as challenging as filming the wind? The last years landscape architecture itself faced new challenges and we found that new landscape-architecture should be more dynamic, more resilient and work more with the forces of urbanisation and nature then against them.

In European landscape-architecture changed quite a lot these last years. For years the field was strongly influenced by the house building boom, and landscape architects became more and more involved in urban design. With many countries within Europe fighting a recession, the home market changed and quite a few European landscape-architecture offices went abroad and continued designing big housing and green urbanisation projects outside of Europe. Within Europe the challenges changed. At least we found the pace of developments slowing down, thus giving more emphasis on the factor time in the design. When the design will not be substantiated within the first decade, one has to alter it, by giving it a temporary, transitional aspect, or organising the design to grow within time. Though we were always aware of this aspect in the design, our surroundings seem to be more and more able to value this aspect. But it is not only a wind less strong, it is a new wind as well.

Within Europe we see the urbanisation continuing. The insight of two themes changed nevertheless: the development of our cities have to become much "greener" than expected, and the rural areas are facing a decline of urbanisation bigger than expected. Both themes give a need and an opportunity for landscape architecture.

A greener urbanisation is not just a trend. Even in these recession years, we have seen that themes as fine-dust, climate change, water-retention, water-management, urban farming, sustainability, CO_2 footprint etcetera became more and more a solid factor in the landscape design. Even though one could have expected them to not be "recession-proof", still we find these theme's here to stay. Thus giving a new, or at least a deeper dimension to the landscape architecture. No longer aesthetical quality is proof enough, the new landscape designer has to find a way to combine the necessary aspects of so many fields into his design. And not only should it fit in, but as well being sustainable for the long run. So we use the green design for the urban climate in terms of control of city temperature or fine dust as well, and find a place in our design for these same plants and trees to contribute to the urban fauna, the well being of the people or even

欧洲景观设计的新风

在生命即将走向尽头之际，丹麦著名的纪录片制作人尤里斯·伊文思拍摄了杰作《风的故事》。在影片中，伊文思来到了中国，试图"拍摄出他穷其一生想要拍摄的风"。这一比喻同样适用于景观设计，因为设计景观的难度几乎无异于拍摄风。近年来，景观设计面临了一些新的挑战，我们发现新的景观设计应当更富活力、更坚韧、更能适应城市和自然环境。

近年来，欧洲景观设计改变了许多。该领域深受房屋建造热潮的影响，而景观设计师也越来越多地参与到城市设计中。由于欧洲许多国家都面临着经济衰退，本土市场变化很大，因此一些欧洲景观设计公司开始向海外拓展市场，在欧洲以外的地区继续设计大型居住区和城市绿化项目。欧洲景观设计正悄然发生变化，发展的脚步已经放缓，更多的注意被放在设计的时效上。如果一个设计不能持续十年的话，就必须被换掉，用临时的过渡设计取代或者是使其持续成长。我们在设计过程中很清楚这一点，而周边的环境也对此越来越重视。这不仅是一股强劲的风，还是一股新风。

在欧洲，我们将城市化看成一个持续的过程，但是有两个情况已经发生了变化：一是城市的发展必须比预期的更"绿色"；二是乡村正面临着比预期更严重的城市化衰退过程。二者都为景观设计带来了新需求和新机遇。

更绿色的城市化进程不仅是一个趋势。即使在经济衰退的这些年里，细粉尘、气候变化、保水性、水处理、城市农业、可持续发展、二氧化碳排放等主题在景观设计中仍然变得越来越重要。尽管一些人曾期望它们不会成为"防衰退"的

make it part of traffic safety. The palette is bold and even more interesting.

Another theme that changed is the decline of urbanisation in some of the rural areas of Europe. For years we have thought, worked, designed, like the urban and more rural areas would develop in the same way. Maybe not in the same pace, but at least the assignment for the landscape architect was more or less similar. By now, we know it is not. Where in the big cities the urbanisation goes forth, as a force of itself we have to manage, in rural areas we have to question the strength of urbanisation. Often we seek for reinforcing the strength of a rural area. For instance redesigning the public area can help a rural place that struggles with living on tourism a lot. It can just be the trigger to get the place going again. Or the need is to find a new way, coping with a declining number of inhabitants. This enhances the scope of landscape architects, or at least should do so. Interesting enough one can see possibilities for increasing wildlife habitat within Europe, where for decades it struggled and seemed not able to co-exist with an industrialised, first world continent. This gives the landscape architecture an assignment on a literally large scale: what do our rural areas look like in the next decades, and how do we want them to be like?

The modern European Landscape architecture is rapidly changing to a more dynamic approach. We realise now that we are part of a fast changing world, with well informed citizens. The awareness that climate change in as well temperature and amount of rain has become an aspect of our work field has sunk in.

We realise now that the effects of urbanisation are more complicated and challenging than we have thought for a long time. A new and much needed wind in European landscape design.

Johan Buwalda
Nijmegen, 17th of April 2015

工具，但是事实却是如此。它们为景观设计带来了新的，或者至少是更深层次的维度。只有"美"已经不够，景观设计师必须将这些必要的主题融入自己的设计。这些主题不仅要契合，还要实现可持续发展。于是，我们开始利用绿色设计来控制城市气温或细粉尘，利用花卉和树木来促进城市动物种群发展、促进人类健康、甚至是保证交通安全。景观设计变得越来越丰富多彩、充满乐趣。

另一方面，欧洲一部分乡村地区的城市化正面临衰退。多年以来，我们一直以同样的方式对城市和乡村进行设计和开发，虽然速度可能有所差异，但是景观设计师的任务一直是类似的。现在，我们才发现做错了。大城市的城市化进程持续向前，而我们对乡村地区的城市化力量却不得不抱有疑问。我们一直尝试强化乡村的力量。例如，公共区域的重新设计能帮助乡村振兴旅游业，它能帮助乡村继续发展。或许我们需要找到一种新的方式来应对日渐减少的乡村人口。这拓展了景观设计师的职责范围。有趣的是，未来欧洲的野生生物栖息地可能会越来越多，这正是多年以来人们求之不得的，因为野生生物栖息地似乎很难与工业化大陆共存。这让景观设计面临更大的任务：我们的乡村在未来几十年会变成什么模样？我们又期望它们变成什么模样？

现代欧洲景观设计正变得越来越有活力。我们认识到自己是飞速变化的世界的一部分，所有公民都能获得丰富的信息。气温、降水量等气候变化已经变成了我们工作所必须面对的一个重要部分。

城市化效应比我们预期的更复杂，也更富挑战性，这就是欧洲景观设计的新风。

约翰·布瓦尔达
2015年4月17日，奈梅亨

Contents 目录

Public Park
公园

- 008 Georgswall
  乔治墙公园
- 014 Recreational Park Berlin-Marzahn
  柏林马尔占休闲公园
- 020 Nature Playground Sloterplas
  斯罗特湖自然游乐场
- 026 Bellamy Park
  贝拉米公园
- 032 Waterwin Park
  水胜公园
- 038 Billie Holiday
  比利假日公园
- 044 The Green Core of Hainholz
  海恩霍兹绿地
- 050 Entranceway, Neue Messe
  汉堡新会展中心入口区域
- 056 Spree Harbour
  斯普雷港口广场
- 062 Fortress Ehrenbreitstein
  艾伦布莱斯坦堡
- 068 Moabiter Stadtgarten
  莫比特城市花园
- 074 Freiaplatz
  弗莱亚广场
- 078 North Park Pulhei
  普尔海姆北方公园
- 082 Park on Harburg Castle Island Hamburg
  哈尔堡城堡岛公园
- 086 Südplateau
  南方高地
- 090 Campa de los Ingleses Park
  英人土地公园
- 096 Lemvig Skatepark
  莱姆维滑板公园
- 102 The Pulse Park
  脉动公园

Urban Public Space
城市公共空间

- 108 Rotterdam Westerkade and Parkkade
  鹿特丹韦斯特码头和公园码头
- 112 Blokhoeve
  布罗克霍温
- 116 Kavel K
  多K青年溜冰运动场
- 122 Laan van Spartaan
  斯巴达大道
- 126 Water Square Benthemplein
  本瑟姆水广场
- 132 Station Area Voorburg
  福尔堡站区景观设计
- 136 The Ravelijn Bridge
  拉维利因堡垒岛之桥
- 140 Hamburg International Garden Show Aqua Soccer
  汉堡国际园艺展水上足球池
- 144 Kala – Playground and Green Space in Berlin-Friedrichshain
  KaLa 弗里德里希斯海因游乐场与绿色空间

148	Saarbrücker Place 萨尔布吕肯广场	208	Biomedical Science Centre 生物医学科学中心
152	Think K K思想广场	212	Leonardo Royal Hotel 莱昂纳多皇家酒店
156	Palmeral El Palmeral of Surprises 帕尔梅拉惊喜长廊	218	Hospital La Paz 拉巴斯医院
160	Garden of the Silhouettes 剪影花园	222	Hospice Djursland 日德兰临终关怀医院
164	Ricard Viñes Square 理查德·韦恩斯广场	228	Aarhus University 奥尔胡斯大学
168	Hyllie Plaza 希里广场		**Private Garden & Residential Landscape** 私人花园与住宅景观
	Commercial & Institutional Space 商业与综合景观	234	Narrow City Garden 窄城花园
174	Talentencampus Venlo 芬洛人才校园	238	Heibloem 喜花农庄
178	International School Eindhoven 艾恩德霍文国际学校	242	Wijsselse Enk 威斯克安科农庄
182	Science Park Amsterdam 阿姆斯特丹科学公园	248	Garden of a Villa in Berlin-Dahlem 柏林 – 达勒姆别墅花园
188	Police Fire Brigade Apeldoorn 阿培尔顿公安消防局	254	1001 Night House 一千零一夜住宅
192	Beukenhorst Zuid Haarlemmermeer 哈勒默梅尔勃肯霍斯特南区	258	House in Somsaguas 索玛古亚斯住宅
196	Square of Knowledge 知识广场	262	The Garden of Water House 水屋花园
200	Student Residence Siegmunds Hof Building 13 齐格蒙德霍夫学生宿舍 13 号楼	266	Casa de Campo 田园之家
204	KPM-Quarter-Plot5 KPM 区第五地块		**Index** 索引

GEORGSWALL
乔治墙公园

Location: Aurich, Germany
Completion: 2014
Design: POLA
Photography: Martin Mai
Area: 16,500sqm

项目地点：德国，奥里希
完成时间：2014年
设计师：POLA景观事务所
摄影：马丁·麦
面积：16,500平方米

The history of Georgswall can be told in many different ways. There is the story of the defense wall to protect the city against the Vikings. Another one is the story of the canal with a harbour, a harbour that was later used for landfill. And there is the story of the Georgswall as a garden strip in the beginning of last century. In the end of the 20th century the gardens turned into a parking area and a place for a weekly market without any sustainable values. This lack of values and identification was about to be solved when a redesign-competition was tendered by the City of Aurich in 2008.

The office of POLA (POETIC LANDSCAPE) was happy to win the competition by convincing the jury with their concept of storytelling design. The idea was to give hints and links to the different historic layers of the Georgswall.

The linear and open dominance of the former defense wall which is set in contrast to the narrow, interwoven streets of Aurich made it possible to visualise a new green belt. Cars were banned and a new area of the market place was found within the Georgswall. By freeing the central green, a new city park was born.

Nowadays four water basins trace the form of the historic harbour. Within the water basins there is a writing saying "OLL HAVEN AUERK", which means old harbour Aurich in an old north German dialect. The 2.5m large letter sculptures are placed underneath the water surface of the basins. The letters are hewn out of huge shell limestones. Due to the size of the stone letters, one can only slowly read the meaning by walking by the basins. The meaning is a link into history, a hint of bygone times. The filled-in harbour, the sunken harbour is now imprinted on the surface of the main plaza. Such hints and phrases are meant to trigger our fantasies.

The Georgswall is in a good way a fragment of larger possibilities, possibilities and beauty that lies in the eyes of the beholder. Phrases perfect themselves and convey its meaning within the realm of our self's.

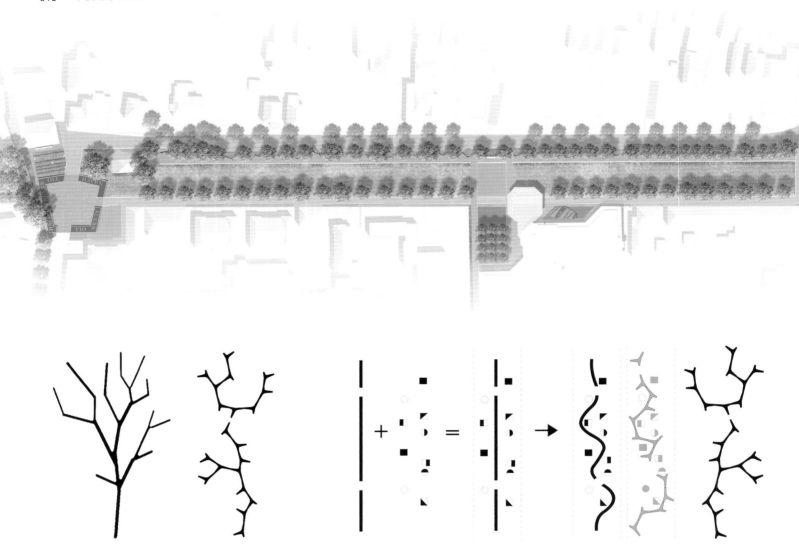

The designer's aim was not reconstruction but identification. It means an identification of the people with the vernacular, the presence the past of the Georgswall. A confusion of time and space to unlock doors in our deepest thoughts.

乔治墙的历史十分丰富。它曾是帮助城市抵御维京人的防御城墙,也曾是一条配有港口的运河,港口后来变成了填埋区。20世纪初期,乔治墙地区被改造成一个园林带。20世纪末,园林又被改造成停车场和礼拜集市,毫无可持续价值。2008年,奥里希市举办了一次针对乔治墙地区的设计竞赛,旨在彻底改变该地区的价值和形象。

POLA景观事务所通过他们的"故事性设计"赢得了评委的青睐,获得了竞赛的优胜。他们的方案是通过提示将乔治墙的历史串联起来。前防御墙宽阔的直线线条与奥里希市狭窄交错的街道形成对比,更像是一条全新的绿色腰带。乔治墙地区禁止汽车通行,并且重新划拨了一块集市场地。设计师通过解放中央的绿地而营造了一个全新的城市公园。

四个水池按照历史港口的位置造型进行布局。水池里写着"OLL HAVEN AUERK",即德国北部方言中的"老奥里希港"。2.5米的字母雕塑被放置在水面之下,由大块的介壳灰岩雕琢而成。由于石头字母巨大,人们只能边走边读出单词的含义。这一设计与历史相联系,让人们体验过去的时光。被填埋的港口如今重现铭刻在主广场的地面上。这种暗示充分触发了我们的想象力。

在旁观者的眼中,乔治墙意味着巨大的机遇,充满了无限的可能和美好。水池里的文字让我们充分感受到了乔治墙的意义。

设计师的目的不是重建,而是塑造一种认同感。这是人对本地文化、与乔治墙的历史的认同感。设计通过时间与空间的总结开启了我们内心深处的大门。

RECREATIONAL PARK BERLIN-MARZAHN
柏林马尔占休闲公园

Location: Berlin, Germany
Completion: 2012
Design: Rehwaldt LA, Dresden
Photography: Rehwaldt LA, Bautzner
Area: 45,000sqm
项目地点：德国，柏林
完成时间：2012年
设计师：Rehwaldt景观建筑事务所
摄影：Rehwaldt景观建筑事务所
面积：45,000平方米

The park was implemented within the concept for the Berlin Garden Show 1987 which was basically inspired by English landscape gardens. Two examples of this design idea are curved paths and large meadows. Within the last decades tree groups have been grown and form a perceptible characteristic space. The gentle topography of the Wuhle valley has been integrated into the open space concept; the mountain Kienberg – although outside the park area – is the park's backdrop. Geometric structured gardens such as the entrance square or Karl-Foerster-Garden were integrated already into the former park to contrast the landscape style through self-contained characteristic features. The idea of aesthetic and thematic "islands" was picked up with the "Gardens of the World" concept in 2001. From now this was the park's specific feature. The different gardens are defined by a concept of cultural authenticity. Each garden is placed in a certain space embedded in the backdrop of the landscape garden. In this way, the park's characteristic is the contrast between vast landscape and small-seized gardens. From the relations between the gardens a second, rather subtle spatial level derived; within the gardens specific cultural identities can be seen.

With the expansion it was possible to offer quiet recreational sites and new attractions for visitors besides already highly frequented areas within the park. While most areas are planned as large meadows the area at Blumberger Damm is developed for infrastructural features such as parking lots, access and logistics. The existing entrance will have to operate for more visitors including tour groups in the future. Therefore, the new visitor centre with restaurant and event site is going to be implemented nearby. The concept for the park's enlargement maintained the existing character as a popular differentiated garden world and has established new attractions such as environmental education and thematic playgrounds at the same time. The spatial idea of a classic landscape park is a design element which is thematically connecting the old and new part. The topography of the area is not orientated by property lines. It is a spacious part of Wuhle valley, a landscape formerly used for agriculture.

	13. Renaissance Garden
	14. South Meadow Park
	15. Playgrounds
	16. South Entrance
1. North Entrance	1. 北入口
2. Oriental Garden	2. 东方花园
3. Balinese Garden	3. 巴厘花园
4. Korean Garden	4. 韩国花园
5. Japanese Garden	5. 日本花园
6. Karl Foerster Garden	6. 卡尔·费斯特花园
7. English Garden	7. 英国花园
8. North Meadow Park	8. 北草场公园
9. Visitor Centre	9. 游客中心
10. Maze	10. 迷宫
11. Christian Garden	11. 基督花园
12. Chinese Garden	12. 中国花园
	13. 文艺复兴花园
	14. 南草场公园
	15. 运动场
	16. 南入口

The path net is designed like an overall layer which is not differentiating between old park and new one. The new designed "belt walk" describes the outline of the new park meadow and is leading through existing tree groups at the same time. Tree groups and solitaires were going to be rearranged in the park in landscape style. Within a first development stage existing aufwuchs (e.g. willows) and temporary plantings from the 1980s (e.g. poplars) were spatially restructured. Subsequent plantings (e.g. oaks, maples, chestnut) complement or replace those tree groups. So, developing an ideal park landscape is not only realised by new plantings but also by long-term redesigns of existing populations. The relation between the original English landscape park and its German interpretation is significant on site. The English landscape park is used as spatial frame in which different cultural interpretations were integrated. This main idea is an expression of a development of the last two centuries which used the English landscape park as generally understandable design language with respect to the history of gardening.

公园设计以1987柏林园林展的设计为基础，以英式景观园林为设计灵感，其中曲折的走道和大型草坪极具代表性。过去的几十年里，树木得到了生长，形成了极富特色的景观空间。乌勒谷舒缓的地势被融入了开放的空间概念中；尽管在公园之外，金伯格山仍为公园提供了美

好的背景。如入口广场或卡尔福斯特花园的几何造型花园已经与公园融为一体，通过自身的特色与景观形成了对比。2001年的世界园林展选择了"主题岛屿"的概念，这同样为公园带来了深刻的烙印。不同的花园被赋予了文化本真的概念。每个花园都嵌入了具有特定景观背景的环境之中。这让公园在宏大的景观和精巧的花园之间形成了对比。花园之间形成了微妙的空间层次，人们可以在花园内感知独特的文化特色。

扩建让公园得以添加宁静的休闲空间和新的景点。大多数区域都被规划成大草坪，而布鲁姆伯格区则被开发成基础设施，如停车场、入口通道、后勤服务区等。未来，原有的入口必须接待更多的游客，包括旅行团。因此，在入口附近将打造全新的游客中心，配有餐厅和活动场地。公园的扩建保留了分化型园林世界的特色，并且打造了环境教育、主题游乐场等新景点。景点景观公园的空间概念在主题上将新旧部分连接了起来。该区域的地形并没有随着界址线而改变，作为乌勒山谷的一部分，它曾是一片农田。道路网络被设计成一个整体，没有新旧空间的区分。新设计的带状散步路将新公园草坪的轮廓勾勒出来，同时也穿过了原有的树林。

公园内的树林和独树都经历了重新布置。在第一开发阶段，种植于20世纪80年代的柳树和临时植栽进行了空间上的重组，被橡树、枫树、胡桃树等补充或替代。打造一个理想的公园不仅需要种植新植物，还要对原有的植物进行重新设计。如何处理初始的英式景观园林和现代的德式园林改造之间关系是重中之重。英式景观园林被用作空间框架，其中融入了各种不同的文化诠释。这一理念反映了过去两个世纪以来英式景观园林在园林界中的重要性，充分尊重了园林设计的历史。

NATURE PLAYGROUND SLOTERPLAS
斯罗特湖自然游乐场

Location: Amsterdam, the Netherlands
Completion: 2012
Design: Carve
In collaboration with: Municipality of Amsterdam, District New West
Photography: Carve
Area: 30,000 sqm

项目地点：荷兰，阿姆斯特丹
完成时间：2012年
设计师：雕刻景观设计事务所
合作：阿姆斯特丹市新西区
摄影：雕刻景观设计事务所
面积：30,000平方米

More and more people live in cities nowadays. The time that city kids can romp around freely, building huts and getting dirty seems to be gone.

It is therefore not surprising that in recent years there is a tendency to create supposedly 'natural' playgrounds, where children are reintroduced to plants, animals, water and adventurous strolling paths. However, a natural play area requires a certain size; a pocketpark has only little chance of survival. The grass is easily trampled to pieces and the water turnes into a mudd pool.

That is why the Nature Playground Sloterpark is such a welcome exception. The area is located in a sheltered corner of the Sloterpark and its size and the presence of an environmental education centre offers the opportunity to make, for once, a real natural playground. As many materials as possible are locally sourced.

The design consists of different parts. The little lake, which is 'feeded' by the water stream, has a very high water quality and acts as the heart of the natural play area. Children can play here in countless ways in and with the water. There are floating platforms, tree trunks and cable ferries, a cable way crossing the water and a little sandy beach. The swamp is a helofytefilter, which cleans the water in a natural way. The swampy area can be experienced on a long floating jetty, which is also accessible for disabled children. In the overgrown area a lookout platform was build around an existing tree. An old dead tree, in another corner of the park, might be the most adventurous climbing parcours of Amsterdam. The tree is wrapped in an innovative netting structure, which enables children to climb up to 8 meters. In the southern part of the area, a playfield for the smallest children was created. Here, a herb garden, a mini labyrinth and a play hill with a waterplayground are situated, of which the kids can influence the water stream. Together with the education centre, which organises 'nature adventures' throughout the year, the nature playground Sloterplas is without a doubt the most adventurous and most sustainable place to horse around in the city.

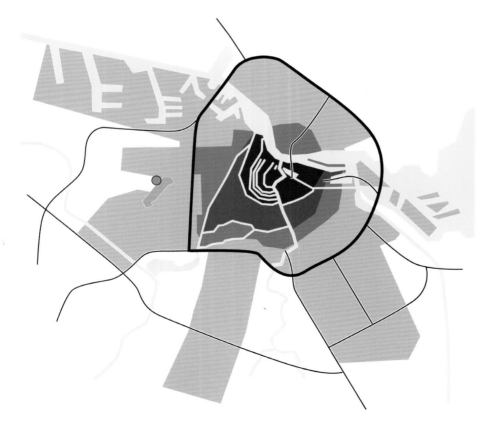

如今,越来越多的人移居城市。从前孩子们在城市中自由玩耍、搭建小屋、弄得全身脏兮兮的时光一去不复返。

因此,近些年来,建造理想中的"自然"游乐场成为了城市建设的一种趋势,在这样的游乐场中,孩子们可以接触植物、动物、水,还喜欢在散步小路上展开冒险。但是,这样的自然游乐场对尺寸有一定的要求;小型的花园很难存活下来。因为草坪容易因人们的踩踏而露出斑斑空地,水池也很快变成泥塘。

这也是为什么斯罗特公园自然游乐场——不同于其他游乐场——如此受欢迎的原因。游乐场位于斯罗特公园隐蔽的中心地带,它的尺寸够大,同时又用作环境教育中心,这两点特质为它提供了机会,使它成为第一个真正的自然游乐场。设计师在为游乐场选择材料时,尽可能多地选用当地的材料。

游乐场的设计由不同的部分组成。一个小湖,由源源不断的水流"哺育",湖水清澈,它是自然游乐场的主体。孩子们能想出数不尽的方法,在湖中或岸

边玩耍戏水。这里有浮动的平台、密集的树干和悬浮的缆车,一条缆车索道穿越水面、跨越小型沙滩。沼泽是一个过滤器,它通过自然的方式净化水质。人们走上长长的浮动码头,便可以近距离体验沼泽区,无障碍通道的设计使残疾孩子也可以进入这里。在植物生长过快的区域、一棵现存树木的旁边,设计师建立了一个监控平台。在公园的另一角,屹立着一棵死亡的老树,它也许是阿姆斯特丹最惊险刺激的攀岩跑酷项目。这棵树被一个新颖的网状结构围绕着,孩子们可以沿着网状结构向上攀爬,高度最高可达八米。在这一区域的南方,设计师为更小的孩子们建设了一个游乐场。游乐场里有香草花园、微型迷宫、游乐小山和戏水游乐园,孩子们可以任意改变游乐园中水流的方向。在教育中心里,全年举办各种"自然探险活动"。斯罗特湖自然游乐场与教育中心组合在一起,无疑成为了这个城市中最惊险刺激且可持续发展的游乐场所。

BELLAMY PARK
贝拉米公园

Location: Vlissingen, the Netherlands
Completion: 2011
Design: OKRA landschapsarchitecten bv
Photography: OKRA landschapsarchitecten, OnSite photography
Area: 25,000sqm

项目地点：荷兰，弗利辛恩
完成时间：2011年
设计师：OKRA景观建筑设计私人有限公司
摄影：OKRA景观建筑设计私人有限公司
面积：25,000平方米

Bellamy Park is located by the harbour of Vlissingen where the Merchant Harbour once stood. Bellamy Park and the Ruyter Square regain a new relationship with the old ports. The Ruyter Square will be slightly raised with sitting stairs at the Fisheries Harbour. The space is organised as a large terraced area, with a slightly recessed centre which rises slowly, with a access for vehicles and an avenue of trees on the west side. The slightly sunken centre offers space for events and doesn't need to be literally green. To make the place also attractive in the evening a lighting plan was made by Guillaume JÈol for the area.

The assignment for the redevelopment of the Bellamy Park, the De Ruyterplein and the Beursplein is the next step in the transformation of the town centre. The Bellamy Park with its surroundings needs to become more than a lively, pleasant and beautiful place to spend time in. The intention is to let it form the heart of Vlissingen's town centre, instead of what it was: a parking square with adjoined greenery.

Hidden harbour – To give the square these functions, the traffic situation is changed from a passage space to a place of residence. The meaning of the former harbour and the meaning of the park in the space will be given a new dimension by creating an elongated middle zone that is just slightly recessed with respect to its surroundings and slowly rises up to the South. It's a poetic translation of what is under the ground and literally cannot be made visible anymore. The edge around the plane is robust, suiting the maritime character of Vlissingen. The park will be bordered by a row of trees and a lane at the West side.

Full and empty, plug and play – The sequence of the intimate middle onto the maritime front requires more than just making the parking square green. The Bellamy Park at the same time is both square and park: due to the facades at three sides of the square and the enclosure of the space, the first impression is that of a square in the town, but the decoration with green and spatial elements lead to the pleasant and relaxed atmosphere that is so typical for a park. The area is lively because of daily activities and the presence of people. The events that regularly take place strengthen this.

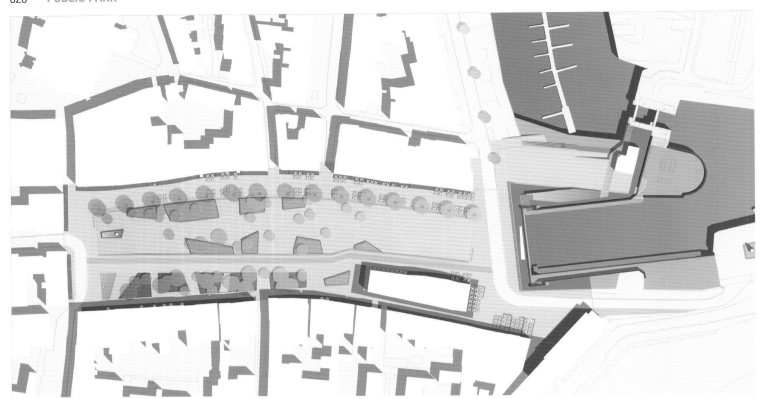
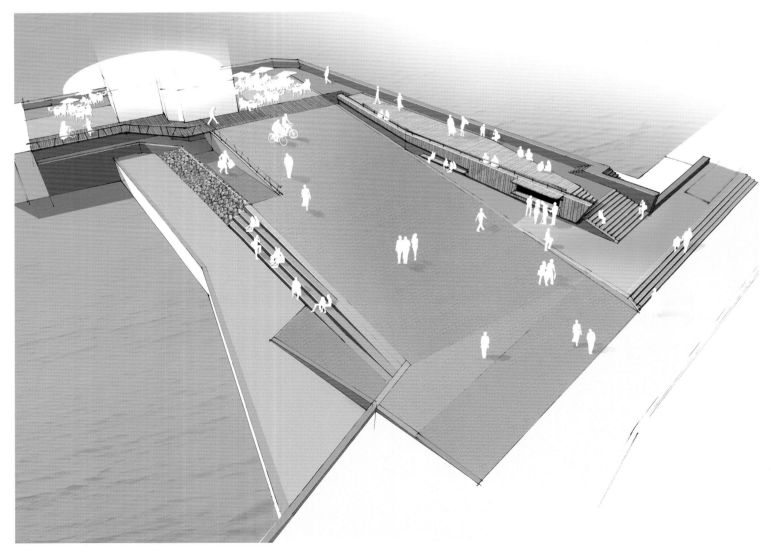

贝拉米公园所在的弗利辛恩港，原来是商业港口。它和勒伊特广场一起，构建起了与老港口间新的联系。勒伊特广场地势稍高，在面向渔业港口的一侧，设有可供坐下休息的台阶。广场如同大型的阶地，中心通过平缓的坡道稍微下沉，西边设有车辆通道和林荫道。下沉广场为人们提供了活动的空间，因此无需绿化。为了使它在夜晚同样具有吸引力，设计师纪尧姆·吉尔为这一区域设计了照明方案。

贝拉米公园、勒伊特广场和比尤斯广场的再开发项目，是市中心区域改建的下一步任务。贝拉米公园及其周边环境需要成为一个更具活力、舒适、美丽的场所，以供人们消遣。设计师希望通过改建，使它成为弗利辛恩城中的核心，而不是原有的带有绿化的停车场。

陆封港：为了赋予广场所需的功能，设计师对这里的交通进行了改造，将原有的通过空间转化为停留空间。为使狭长的中心与周围环境相协调，这一区域稍微下沉并向南方缓慢上升。通过这样的改造，前港口和公园被赋予了新一维度的意义。这是对地下不可见的事物诗一般的诠释。广场的边缘线条粗犷，反映出弗利辛恩市沿海的特征。公园的西侧，一排树和一条车道界定出它的边界。

虚实相应，即插即用：从中央区域的隐蔽到沿海滨岸的开放，空间序列的变化，需要的不只是简单的停车场绿化。贝拉米是一个广场，同时也是一个公园：由于三侧建筑的立面形成的围合空间，它给人的第一印象是一个城市广场，但是植物和空间元素的装饰所形成的愉悦和悠然的气氛，却是公园的典型特征。由于日常活动和人流的往来，这里充满活力。定期举办的大型活动也为其增加了生气。

030 • PUBLIC PARK

WATERWIN PARK
水胜公园

Location: Zeist, the Netherlands
Completion: 2011
Design: OKRA landschapsarchitecten bv
Photography: OKRA landschapsarchitecten bv
Area: 20,000sqm

项目地点：荷兰，泽斯特
完成时间：2011年
设计师：OKRA景观建筑设计私人有限公司
摄影：OKRA景观建筑设计私人有限公司
面积：20,000平方米

Because the function of water extraction the green area on the corner of Sumatra Avenue and Mountain Road to date has not been built. OKRA created the design in an interactive process with local residents of the park. For this park is very literally: without water no park! After all, without the function of abstraction, this space within the core of Zeist may not even be a park.

A water playground makes the function of abstraction for children visible and gives a possiblility to experience. In the floor of the promenade, a cartridge was made that refers to the filtering of the water in the soil. The water is pumped up from approximately 80 meters deep. The installations of Vitens get a prominent place in the park.

The park is an open space in the woods. The existing trees surrounding the open space remain. Besides the existing windbreaks some tree groups are added.

In the open space several nuances are used. Tightly mowed grass, more rugged grass and a meadow of natural flowers alternate.

From the edges of the park views through the park are offered. The view towards the school remains and the see through towards the towers of the Vitens pump and the gym are highlighted.

To the existing round through the park a number of routes are added in order that you can make several laps through the park.

The three edges of the open space in the woods have different atmospheres. The forest atmosphere on the western edge is broadened and here is room for games in the woods. The high bushes along the Mountain Road will be replaced by lower bush plants with more flowering vegetation. When you pass the school you see space for school gardens and playgrounds for the smaller children.

公园 • 033

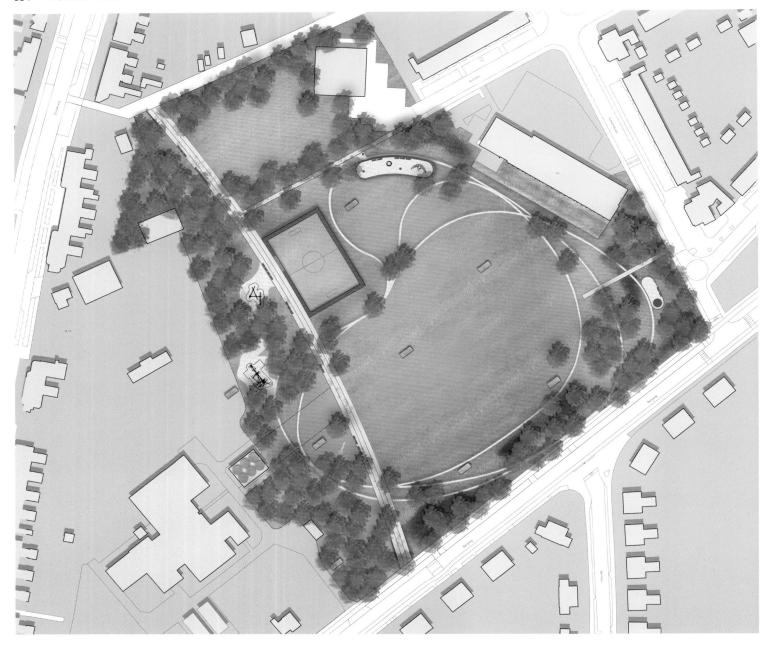

The park offers new program for everyone. By working in a clever way with the programming of the space it is possible to give space to all different groups.

由于水提取功能，苏门答腊街和山脉路交界处的绿化区域至今还未建成。OKRA景观建筑设计公司创新性地采用了交互式设计程序，在设计中和公园当地的居民进行互动。这个公园的特征，就如同它的名字所诠释的：没有水，没有公园！毕竟，没有集水的功能，这个位于泽斯特核心的空间甚至不能称作是一个公园。

戏水游乐场将集水的功能展现给孩子们，并使他们有机会亲身体验。在散步道的地面，设计师装了一个筒状物，用于过滤土壤中的水份。水通过水泵从大约80米深的地下抽取上来。维坦斯水泵安装于公园的显著位置。

公园位于森林中的一片开阔空地。设计师保留了空地周围原有的树木，并在现有的防风林旁增加了一些树木群落。

林中空地有一些细微的差别：整齐修剪的草地、更加自由生长的草坪，以及开有野生花卉的草坪。

站在公园的边缘，整个公园尽收眼底。朝向学校的景观被保留了下来，穿越维坦斯水泵塔楼和体育馆的景观视线被特别强调。

在公园原有的环形路上，增加了多条路径，这样人们游览公园时有多种路线可供选择。

森林空地的三个边缘有着不同的氛围。在西侧边缘，森林的环境特征被放大，设计师在那里布置了林中游戏的空间。沿着山脉路的边缘，高的灌木丛将被替换成更多低矮的开花灌木植物。靠近学校的区域，有校园花园和供小一点的孩子使用的操场。

这个公园为每个使用者都提供了新的使用方案。通过对空间进行灵巧的设计，设计师为所有不同的群体都提供了各自的空间。

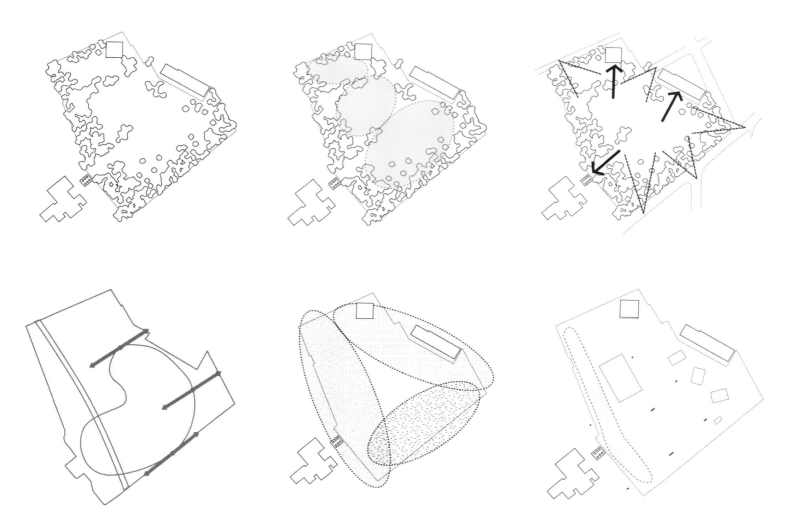

BILLIE HOLIDAY
比利假日公园

Location: The Hague, the Netherlands
Completion date: 2013
Design: Carve
Photography: Carve
Area: 2,000sqm

项目地点：荷兰，海牙
完成时间：2013
设计师：雕刻景观设计事务所
摄影：雕刻景观设计事务所
面积：2,000平方米

The Billie Holiday Park is situated in Loosduinen, a quiet suburb of The Hague. After a recent renovation of the park, this part seemed to be forgotten; it was an unused, neglected space. However, this changed when a residential-care complex was built. Suddenly, this part became a frontside, which created the possibility to transform this place into an attractive park zone, too. The community board organised a participation process, mapping the wishes of the local residents. Renovating the soccer field and guarding off the biking path were important starting points, just like improving the spatial structure of the park. By arranging all new functions in one binding element, the surrounding space is reorganised, too.

Carve designed an organically shaped playhill with three 'heads', which curls around an existing tree like a stretched piece of elastic. Because of this addition the surroundings are redefined, focused on its neighbouring functions like the residential care complex and the houses. The edges of the object continuously transform; from a sitting edge, tribune and treebench to a two meter high climbing wall. The two lower edges are designed as a platform, in which various playelements are integrated. The highest bulge surrounds a sheltered space with more dynamic elements.

The sculpture merges, because of its fluid forms and continuous skin, into one large playobject which attracts all ages and ability levels. Because of its shape and colour the object is an important functional addition to this part of the park, which in a short time span has become the new meeting place for the whole neighbourhood.

比利假日公园地处于海牙静谧的郊区洛斯德伊嫩。在公园最近一次的翻修之后，这里一直处于闲置状态，未被利用，它似乎已经被人们遗忘。但是家庭护理综合体建筑建成后，情况发生了改变。一夜之间，这里成为街区的正面，也获得了转变为具有吸引力的公园的机会。社区委员会组织了一次分享会，收集了当地居民们的愿望。翻新这里的足球场、建设自行车道保护设施是重要的开始，同样的改建措施还包括改善公园的空间结构。设计师将所有新功能整合进一个连接元素，也通过这种方式把周围的空间重新组织起来。

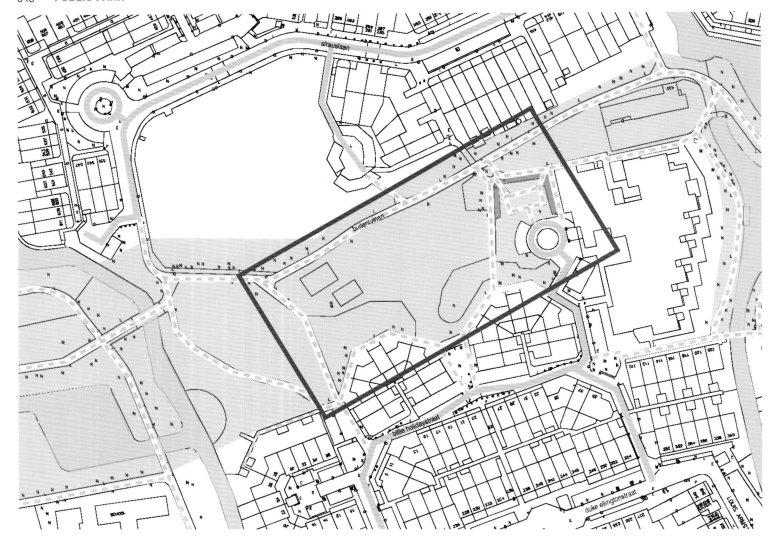

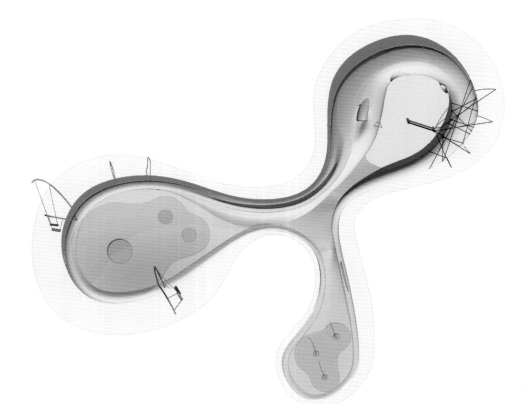

雕刻景观设计事务所设计了一座有机体外形的游乐小山，小山有三个"山头"，像一条拉伸的松紧带，盘绕在一棵保留树木的周围。小山的添加，重新定义了周边环境，使邻近区域，例如家庭护理综合体和住宅的功能得到了强化。小山的边缘连续不断地变化，从可以坐人的边沿、廊台、树木长椅，到一座两米高的攀岩墙，高低起伏。两个较低的边缘被设计成平台，平台上集合了多种多样的娱乐设施。最高的山丘围绕着一个隐蔽的空间，那里有更多的动态元素。

作为一个巨大的游乐设施，这座小山流动的形状和连续的表皮，吸引了不同年龄、不同能力水平的人群。由于其突出的形状和颜色，它成为公园这一区域重要的功能性附属设施。在短时间内，这里便成为所有社区居民新的集会场所。

THE GREEN CORE OF HAINHOLZ
海恩霍兹绿地

Location: Hanover, Germany
Completion: 2012
Design: Büro Grün plan
Photography: Thomas Langreder, Verena Oesterlein
Area: 26,200sqm

项目地点：德国，汉诺威
完成时间：2012年
设计师：Büro Grün规划事务所
摄影：托马斯·兰格里德、维利纳·奥斯特林
面积：26,200平方米

The "Green Core" in Hanover's Hainholz was planned as a local "Park of the Generations" with a multitude of possibilities for visitors of all ages. It is part of the wider "New Centre" regeneration in this part of Hanover. Its aim is to further improve social and cultural cohesion in a visionary, lasting, town development. An array of public services, with Cultural Centre, Family Centre and sports facilities are all linked within this open space of 3 acres.

The planning focus gravitates towards the vibrant, urban square at the foot of the Cultural Centre, and the Family Centre in the southern corner of the park. Old trees, now amidst a lush expanse of grass are accentuated and framed by a structural semicircle set in stone. Furthermore the "Bank of the Generations" follows and copies the flow of the landscape and provides a meeting point for communication, play and activity.

Between the Cultural Centre and the Family Centre, the linear "Play Path" links the urban area with the northern open green landscaped garden. There are many play opportunities for all ages, from sand pits for toddlers to a playground for children, as well as a football pitch and a basketball court. They all offer a wide range of activities. The fitness point is aimed at the senior visitor with a selection of training and exercise installations which prove to be popular. These facilities are grouped together and were carefully chosen having consulted the relevant user group during the planning process.

The green, open space in the northern half of the park is dominated by an everchanging, rolling landscape, even including a tobogganing slope. The old body of trees has been diversified by planting additional, younger trees among them.

The new "Green Core" is embedded in the local network of walkways and cyclepaths, east-west and north-south. From the square in the southern corner two paths head northwards, the main one linking it to the entrance of the "Hainholz

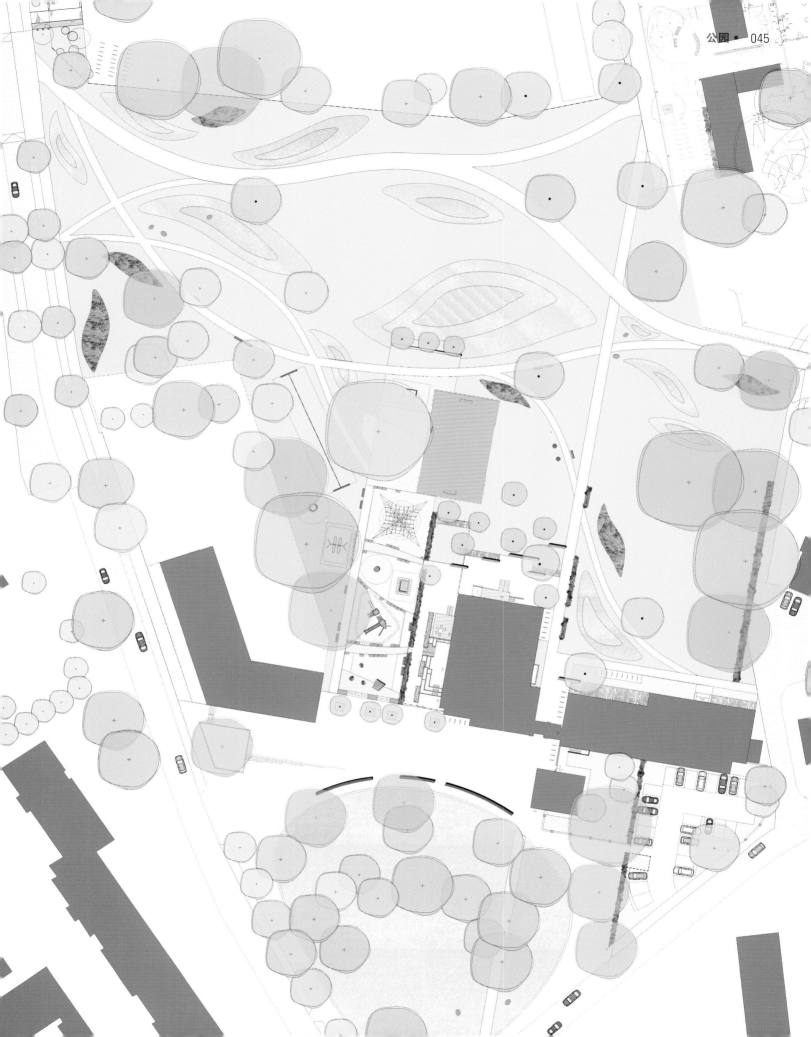

lido". The windy "Julius-Trip-Ring" in an east-west orientated axis connects the "Hainhölzer Market" to adjacent allotments residential areas and beyond.

汉诺威的海恩霍兹绿地被规划为当地的"世代公园",各个年龄段的游客都能自得其乐。作为汉诺威"新中心"重建计划的一部分,它的目标是通过有远见的城市开发来改善社交与文化环境。这块占地26,200平方米的开放空间融入了一系列公共服务,包括文化中心、家庭中心、运动设施等。

项目将重点放在城市广场上,文化中心设在广场地势上方,而家庭中心则设在公园的南角。被青草包围的古树外围圈起了半圆形的矮石围墙。"世代河岸"随着景观环境而变化,为人们提供了交流、游戏和活动的集会场所。

在文化中心和家庭中心之间的"游戏路"将城市区域与北面开阔的绿地景观花园连接起来。绿地设有多种老少咸宜的游戏设施,从幼儿沙坑、儿童游乐场到足球场和篮球场。健身点专为老年人所设计,配有一系列广受好评的健身设施。这些设施聚集在一起,设计师在规划阶段分别详细地咨询了对应的用户群体。

公园北半边的绿地空间是不断变化、高低起伏的景观地势,甚至还包含一个雪橇坡。各种各样的新栽树木将古树环绕起来,形成了多姿多彩的景观。

新绿地嵌入了当地四通八达的人行道和自行车道网络。两条道路从南角的广场一路向北,主路直达海恩霍兹海水浴场的入口。"朱利叶斯旅行圈"是一条东西向轴线,将海恩霍兹市场与相连的居民区连接起来。

ENTRANCEWAY, NEUE MESSE
汉堡新会展中心入口区域

Location: Hamburg, Germany
Completion: 2011
Design: A24 Landschaft
Photography: Hanns Joosten

项目地点：德国，汉堡
完成时间：2011年
设计师：A24景观事务所
摄影：汉斯·尤斯登

In the city of Hamburg, between the Elbe River and Alster and opposite Hamburg Messe, the city's exhibition centre, lies Planten un Blomen – one of Hamburg's most beloved parks, a park that enjoys an eventful history. As part of an urban reorientation and the expansion of the adjoining trade-fair grounds, the park entrance just across the street was redesigned by A24 Landschaft in 2011.

A new fence surrounds the southern grounds of the park like a sash. It consists of steel plates that play with light and shade, that delimit and protect, yet allow the gaze to penetrate and provoke curiosity; for, from the outside they look deformed, as if the nature contained within the park is protesting its confinement, trying to escape.

The design confronts the ambivalence of protecting the precious park in the middle of the metropolis, while insistently beckoning visitors to a new ramp, which re-establishes the entrance to this side of the park as a clean incision. It breaks through the thick green façade of the old stock of trees, managing simultaneously to preserve this façade and to direct people's gaze into the park. Both careful interventions create a new spatial dramaturgy between movement and rigidity, between city, nature and culture.

在汉堡市易北河与阿尔斯特河之间，汉堡会展中心对面，坐落着深受汉堡人所喜爱的花卉植物园。这座植物园的历史十分丰富。作为城市重新定位和会展地块扩建的一部分，A24景观事务所于2011年对对街的公园入口进行了重新设计。

新加的栅栏像腰带一样将公园南面的地块环绕起来。它由光影交错的钢板构成，一方面将公园保护起来，形成边界；另一方面让人可以瞥到公园的一角，吸引着人们。从外面看，栅栏是变形扭曲的，好像公园里的自然植物被监禁起来，正试图逃跑。

设计一方面保护着城市中心珍贵的公园，一方面又通过新建的坡道吸引游客，在公园的一侧重塑了简洁清晰的入口。它穿过厚厚的树丛，既保护了绿色空间，又成功地将人们的视线吸引到公园。精巧的设计在运动与静止，城市、自然与文化之间形成了全新的空间关系。

SPREE HARBOUR PLAZA
斯普雷港口广场

Location: Hamburg, Germany
Completion: 2013
Design: TOPOTEK 1
Photography: Hanns Joosten
Area: 28,838sqm

项目地点：德国，汉堡
完成时间：2013年
设计师：TOPOTEK 1景观事务所
摄影：汉斯·尤斯登
面积：28,838平方米

The project is located on the south side of the highly-driven Hafenrand Street, between the Spree Harbour and the residential area of the Reiherstieg quarter, part of an extensive green belt: the Veringpark. The crossing point, west of the Ernst-August-Canal, will be faced with a new function in the future, with which the site's infrastructure, until this point, was not aligned with. Here, the urban aspect of the planning overlaps the transport structure of the Hafenrand street. For easy accessibility to the Spree Harbour expanse, the plans will be accompanied by the Klütjenfelder main embankment crossing plans. There are three locations with stairs leading over the embankment and two locations which enable cyclists to cross.

The Spreehafenplatz is the central square where all paths meet and a special spot for Wilhelmsburg. The fairground captivates viewers through openness and emptiness. The character is to be preserved, it is neither an urban nor a nature spot. Through the specific use of two architecturally placed rows of benches, the quality of the public realm is directly associated with the spot. At this specific place, there is a particular bench in one of the commonly contrary colours: bright yellow. This particular type of bench, adapted to its space through steelpipe-wooden construction, has the character of a leisure and picnic bench. The extraordinary part is the alternating back, allowing a two way view: down to the Spreehafen or around the area and out to the grass steps.

项目位于哈芬兰德街南面，斯普雷港和雷尔斯蒂格居住区之间，是维灵公园绿化带的一部分。未来，埃尔恩斯特运河西面的交叉点将被赋予全新的功能，与场地现有的基础设施并不匹配。城市规划将哈芬兰德街的交通结构叠加起来。为了方便进入斯普雷港，设计将配有堤岸交叉规划：共有三组台阶跨过堤岸，其中两组可供自行车通过。

斯普雷港口广场是所有路线的交汇点，也是威廉斯堡地区的特殊景点。广场以开阔的姿态吸引着游客，既不是自然景观，又不是城市景观。两排长椅的设计赋予了广场公共空间的品质。色彩亮丽的黄色长椅由钢管和木材制成，极富休闲之感，吸引着人们坐下来野餐。长椅的两面分别享有斯普雷港口和草坪阶梯的景色。

058 • PUBLIC PARK

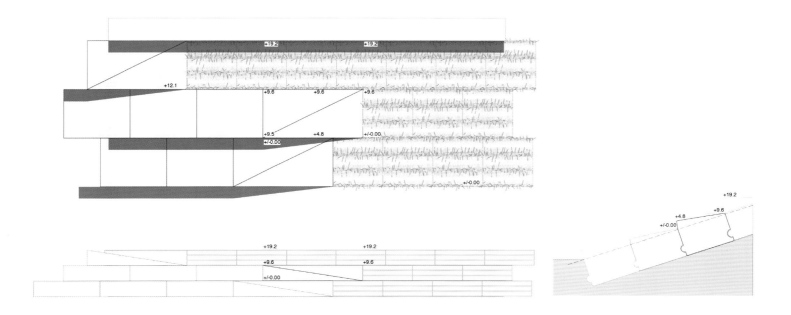

FORTRESS EHRENBREITSTEIN
艾伦布莱斯坦堡

Location: Koblenz, Germany
Completion: 2012
Design: Topotek 1
Area: 95,000sqm

项目地点：德国，科布伦茨
完成时间：2012年
设计师：Topotek 1景观事务所
面积：95,000平方米

The "Festung Ehrebreitstein" (Ehrenbreitstein Fortress), one of Koblenz's landmarks and second-largest preserved fortress in Europe, looks down on the confluence of the Rhine and Moselle Rivers at the "Deutsches Eck" (German Corner) and back on an eventful history.

"As early as five thousand years ago humans settled on the Ehrenbreitstein plateau" explains Romy Zahren of the Koblenz Tourism Board. Until the emergence of the Prussian bastion that can be seen today, different buildings were set on top of each other as explained by Zahren. These layers can be discovered with the start of the BUGA in a very special way: "A glass elevator descends into the origins of the history of the fort".

The impressive stronghold, overlooking the confluence of Rhine and Mosel rivers is an important national monument. In a comprehensive restructuring of the northern approach as a visitor's facility, the historical plateau is developed as a museum park and the spatial qualities of the site brought into context with the wider surrounding.

The general design concept reorganises the historic fortress plateau in a spatial and dramaturgic way into the new main entrance of the Fortress Ehrenbreitstein. The car entrance and the parking lot are placed at the edge of the plateau. Through this the big space keeps is monumentality and stages the silhouette of the Fortress to the north. On top of the former baroque design construction is a wide grass field laid out which makes out the background for the experience of the Fortress ensemble. A new network of path axes orders the plain and is connected with the existing system of pathways.

艾伦布莱斯坦堡是科布伦茨的地标性建筑，也是欧洲现存的堡垒中第二大堡。它俯瞰着莱茵河与摩泽尔河的交汇处，拥有丰富的历史。

科布伦茨旅游局的罗米·扎兰介绍道："早在5000多年前，人类就已经开始在艾伦布莱斯坦高原上定居了。"直至我们仍可见到的普鲁士堡垒出现之后，各种建筑开始陆续出现。这些建筑的废墟叠加出不同的层次，人们可以通过新设的玻璃电梯逐渐向下探索堡垒的历史。

这座俯瞰莱茵河与摩泽尔河汇流的宏伟堡垒是极为重要的国家历史文物。在堡垒北侧游客设施的综合翻新工程中，历史高原被改造成博物馆公园，为整个场地的周边环境带来了新的空间特色。

整体设计概念通过空间塑造手法将历史堡垒高原塑造成艾伦布莱斯坦堡的主入口。车辆入口和停车场被安置在高原的边缘。这种设计保持了大空间的宏伟感，在北侧显现出堡垒的剪影。在之前的巴洛克式设计结构之上，设计师构造了大面积的草坪，为堡垒建筑群提供了独特的背景。新增的道路网络与原有的道路系统结合起来，四通八达。

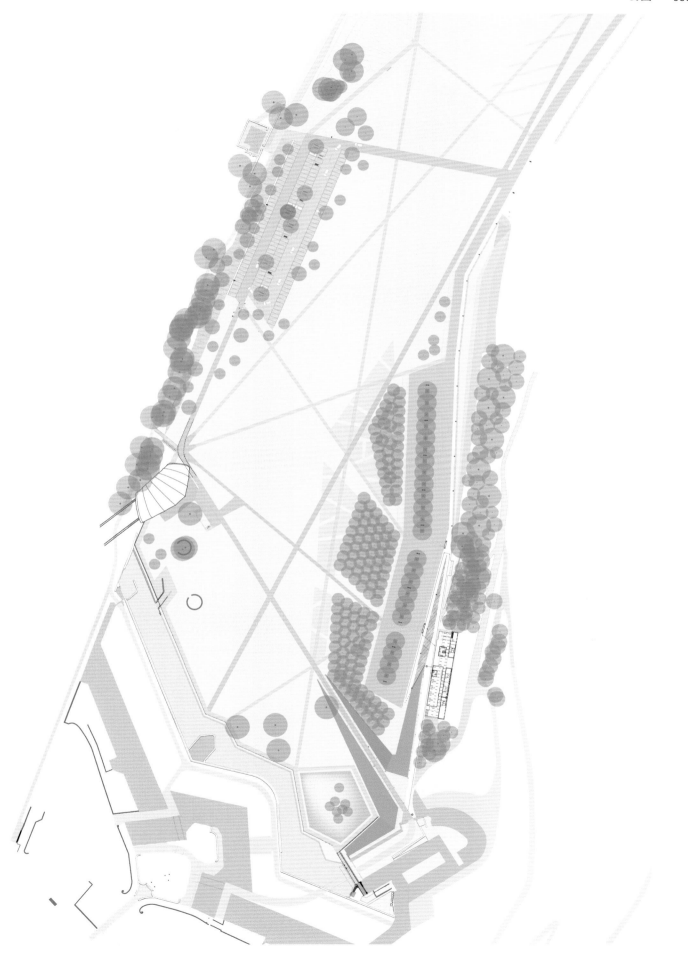

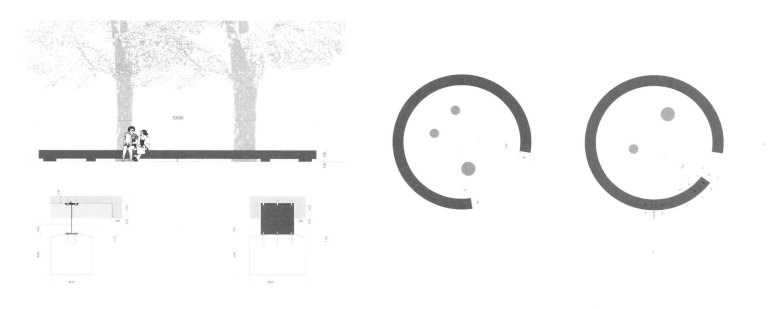

MOABITER STADTGARTEN
莫比特城市花园

Location: Berlin, Germany
Completion: 2012
Design: glaßer und dagenbach, landscape architects bdla
Photography: Udo Dagenbach
Area: 13,000sqm

项目地点：德国，柏林
完成时间：2012年
设计师：glaßer und dagenbach景观建筑事务所
摄影：乌多·达根巴赫
面积：13,000平方米

The basic idea of the concept is an urban garden, in which the residents receive a useful platform as large as possible to develop their activities, by using a clear, easily understandable outline. The park provides services as well for families as for seniors.

The free space in front of the former freight depot Moabit offers the possibility for various uses, for example a space for art and culture, a location for events, a playground or a meeting point in the cosy beer garden. A curved path with seatings leads to the building and to the rear areas of the park. On either side of the forecourt are created lawns under a bright canopy of 100 Japanese pagoda trees (Sophora japonica). Their crowns are cut flat, like a roof.

The area east of the building shall be provided to the public for an individual use. In a collaborative development process, the split area could develop, for example, into a mosaic of kitchen garden, school-based "laboratory garden" and cosy meeting place. The citizen garden receives raised planting beds and structuring rank elements.

On the west side of the building is the playground with various play and activity opportunities for children and adults. For the design of the playground children and young people between 4 and about 10 to 12 years were surveyed. Many wishes could be considered, such as a wave path for bicycles, crates, towers of wooden boxes that point to the former use of the freight depot, partner swings, nest swings and a bobby car race track.

At the north side of the park, the terrain is banked slightly rising so that a "balcony" is created. Sloping to the south a generous lawn/wildflower meadow arises as an orchard with fruit trees and shrubs, which proceeds shall be used by the residents.

The noise-protection wall to the neighbouring street is designed as a concave concrete wall with a structure like "egg-coal" – a reference to the use of the site as a freight depot.

公园 • 069

设计的基本概念是打造一座城市花园，通过简洁清晰的设计为居民提供尽可能大的活动平台。公园的服务应当老少咸宜。

莫比特货运站原址前方的自由空间可以实现多样化的用途，例如文化艺术空间、活动场所、游乐场或休闲的啤酒花园等。一条蜿蜒的小路上布置着座椅，通往货运站和后面的公园。前院的另一侧是草坪和100棵日本槐树，它们的树冠被修剪成平屋顶的造型。

建筑东侧将为公众提供私人使用空间。在协同发展过程中，分隔的区域可以被开发成菜园、学校试验花园以及休闲聚会场所的混合体。市民花园拥有架高的种植池和结构排列元件。

建筑西侧是可供男女老少尽情游戏的游乐场。在游乐场的设计中，设计师对4至12岁之间的儿童和青少年进行了问卷调查。设计综合了他们的诸多愿望，例如自行车坡道、板条箱、木盒子塔（暗指货运站）、伙伴秋千、鸟巢秋千以及警车赛道。

公园北侧的地势稍高，形成了一个"阳台"。南侧缓坡上是宽阔的草坪和野花草地，还栽种着果树和灌木，像一个果园。果园的收益将归居民所有。

临街的减噪墙被设计成凹面混凝土墙，其纹理类似"蛋形煤块"，进一步暗示了地块作为货运站的历史。

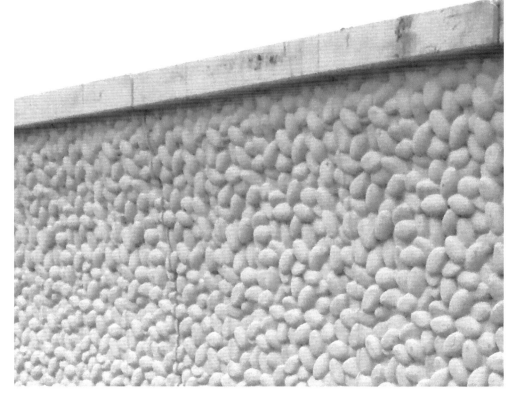

FREIAPLATZ
弗莱亚广场

Location: Berlin, Germany
Completion: 2013
Design: gruppe F
Photography: gruppe F
Area: 2,500sqm

项目地点：德国，柏林
完成时间：2013年
设计师：gruppe F设计事务所
摄影：gruppe F设计事务所
面积：2,500平方米

Improving the Freiaplatz in 2013 was funded as part of the Berlin's urban development programme for the east. As a result of the children participation and public participation process, it focused on strengthening the green, ample character and redesigning playground areas for younger and older children. "Playing under tree canopies" was brought up as one particular idea during the children participation. An existing wall feature was opened up to enable children to "play through" it, three connections to climbing elements were attached at its back and at the sides to create a playing experience above the ground closer to the tree canopies. In the new play area for toddlers, a comfortably sized sitting element for parents completes the sitting and playground area in the shade of the big chestnut. Now this north-western part of the square can be used by all generations.

Rather unappealing elements like the big metal pavilion and the former play area for young children were removed and converted into an ample central lawn for playing, sunbathing or having a picnic. A small hill at its centre alludes to the district's name ("Lichtenberg" translates to "light hill"). A wide perennial planting marks the edge of the central lawn and points towards other interesting features of the square.

2013年弗莱亚广场改造是柏林东部城市开发项目的一部分。广场改造以儿童参与和公共参与为核心，将重点放在加强绿化设施、重新为青少年设计游乐区上。"在树冠下玩耍"是专为儿童参与所提出的概念。设计师特别开放了一面墙景供孩子们游戏，他们在墙的后面和侧面添加了攀爬构件，让孩子们离开地面、更加贴近树冠。在幼儿游乐区，设计师为家长提供了舒适的座椅，让他们可以坐在胡桃树下看护孩子。完善的景观设计让各个年龄段的人群都能在广场的西北区域获得乐趣。

设计师并没有添加任何大型铁亭式的景观元素，并且拆除了原有的儿童游乐区，用茂密的中央草坪作为替代。人们可以在草坪上玩耍、享受日光浴或进行野餐。草坪中央的小山丘暗示着街区的名称——光明山。草坪边界种植着多年生植物，指向广场别处有趣的景观设施。

1. Salon
2. Toddler Area
3. Sand Area
4. Game Lawn
5. Meadow

1. 沙龙
2. 幼儿区
3. 沙坑
4. 游戏草坪
5. 草地

NORTH PARK PULHEIM
普尔海姆北方公园

Location: Pulheim, Germany
Completion: 2012
Design: bbzl böhm benfer zahiri landscapes urban design
Photography: bbzl böhm benfer zahiri, Holtschneider&Peetz

项目地点：德国，普尔海姆
完成时间：2012年
设计师：bbzl景观事务所
摄影：bbzl景观事务所

The park is located between Pulheim's northern limits and the surrounding agricultural landscape. Dirt paths, tree-lined alleys and parceled lots mark the characteristic elements of the meadow landscape. In the transition between city and countryside, the expansive view of the adjoining fields is particularly impressive. Its spatial character is defined by staggered rows of trees, scattered woodlands and farms, and industrial buildings along the horizon.

Structure
By the year 2030, the city limits of Pulheim are to be encompassed by an elongated park on the northwestern edge. It is assumed that over a period of approximately 20 years, the further development of the park will be subject to a number of unforeseeable factors.

The concept of the park is thus intended to allow for adjustments and alterations while retaining a recognisable overall design. The designers propose a spatial network of roads and meadows, which makes use of existing dirt paths, alleys and parceled fields. The public areas of the park include the main paths, splendid views and spatial relationships. They mediate between urban structures and expansive views, and designate the characteristic features of the landscape. Depending on the different needs and initiatives of the surrounding community, the remaining areas may be additionally "filled" with diverse functions over time, operating as semi-private outdoor areas that fluctuate between extensive and intensive use.

Openness
The proposed design concept deliberately disregards a fixed allocation of areas. The spatial elements of the design focus on preserving and strengthening existing scenic qualities. In contrast, the temporal elements of the design allow for adjustments and transformations over an extended period of time.

公园位于普尔海姆北部边界与农田景观之间。泥土小径、树荫小巷和农田地块是草地景观的特色元素。在城市和乡村的过渡区域，辽阔的农田景观十分壮观。整个空间装点着错列成排的树木、分散的林地和农场以及天际线处的工业建筑。

结构

到2030年，普尔海姆的城市边界将由西北部的带状公园所包围。预计在20年的时间内，公园的未来开发将由一系列不可预知的因素所影响。

公园的设计目标是在保持标志性整体设计的前提下，进行调整和改建。设计师提出了一个由道路和草地构成的空间网络，充分利用了土路、小径和农田地块。公园的公共区域包括主路径、优美的景观以及空间联系。它们在城市结构和辽阔的视野之间起到了缓冲作用，奠定了景观的基本特色。根据周边社区的不同需求和规划，公园的其他区域可能会添加不同的功能设施，形成兼具广泛性和集中性的半私人露天区域。

开放性

项目设计有意地忽视了区域的固定配置。设计的空间元素将重点放在保护和强化已有的景观品质上。在未来，设计元素将会随着时间的变化进行调整和转型。

PARK ON HARBURG CASTLE ISLAND HAMBURG

哈尔堡城堡岛公园

Location: Hamburg, Germany
Completion: 2011
Design: Hager Partner AG
Photography: Hager Partner AG
Area: 26,000sqm
项目地点：德国，汉堡
完成时间：2011年
设计师：Hager事务所
摄影：Hager事务所
面积：26,000平方米

Over the millennia, the fortified island rose over the marsh landscape on a slightly elevated site. Layer after layer were added and altered, and uses intersected and overlapped to become a mosaic of fragments that no longer allows the former meaning to be recognised. Nonetheless, the gentle topographic elevation that still distinguishes the site of the castle today does not escape the attentive observer. The castle was pivotal for the history of the location and still allows its former significance to be imagined. This fine topographical feature will be accentuated and further developed as a clearing within a grove of high-stemmed ash trees. This provides the former castle with a representative framing and points to the historical significance of the location. The framework of trees and the clearing emphasise the stellate form of the park, and, along with the construction of new buildings as "bastions" at the vertices, make reference to the former fortress. As a new layer designed in a contemporary manner, the robust concept allows for future developments and facilitates the integration of any historical relicts such as surfacing materials or masonry elements that might come to light.

Harburg Castle: the Centre of the Park

One of the important core features of the park's design is the historic Harburg Castle, located on a slight hill. By making this building, which is steeped in history, the centre of the park, the spotlight has been restored to the old nucleus of Harburg. A large children's play park is being built right beside the historic castle. Children will have a choice between an area where they can run around freely, and a slide, climbing ropes, and hammocks. The play park is well protected by the surrounding trees and offers cool shade in the summer.

Water: a Major Component of the Design

Along with the central position given to Harburg Castle, water also plays a very important role in the park's design. All of the lawns have been designed as clear, open spaces, so that visitors can enjoy unique, unobstructed views of the surrounding waters of the inland port. Visitors and residents can make use of the three points of water access, enjoying the sunshine sitting on steps and jetties,

or relaxing, dangling their legs, and watching the hustle and bustle on the water. Since the waters of the upriver port are not tidal, visitors to the park can sit right at the water's edge.

Trees and Shrubs

Around the park, flowerbeds with shrubs and groups of elms, alders, and willows create a transitional area between the park and the residential development, and thus screen the areas of public and private land from one another. In summer the trees cast refreshing shade, while in autumn they bathe the park in a lush yellow glow. Along the waterfront, shrubs are constantly in bloom, providing a clear demarcation between the park and the water. For residents and visitors to the Schlossinsel, the park provides a fascinating blend of the historical and the modern, a mix that is set to define life on the island.

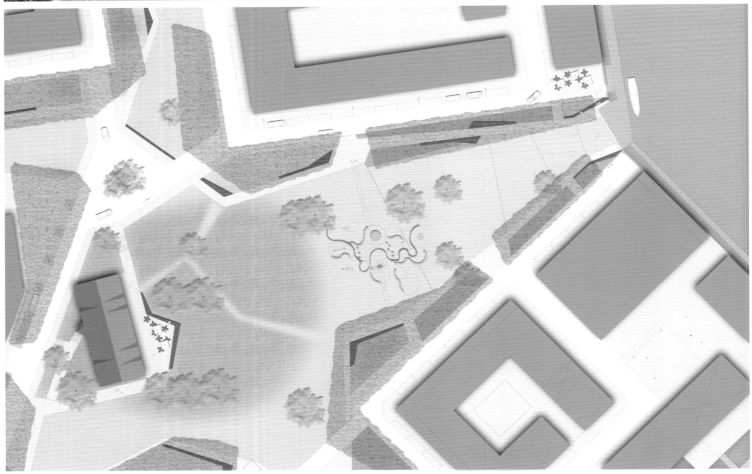

城堡岛从湿地景观中缓缓升起，层层叠加，各种功能相互交错重叠，形成了碎片的组合，已经不再是一个堡垒。尽管如此，略高的地势仍然使城堡所在的地块引人注目。城堡的历史地理位置十分关键，让人无法忽略它的重要意义。优美的地貌特征被进一步开发成白蜡树林中的空地，这为前城堡提供了具有代表性的框架，突出了它的历史意义。树木与空地的框架突出了公园的星状造型，与高处新建的堡垒式建筑共同效仿了之前的堡垒。作为现代设计的新层次，稳健的设计概念有利于未来的开发，并且让残存的历史文物（如墙面材料或砖石结构）得以重见光明。

哈尔堡：公园的核心

公园设计最重要的核心之一就是小山上的古哈尔堡。为了重现位于公园中心的哈尔堡，设计师对原哈尔堡的内核进行了修复。古堡旁建造了一个大型儿童游乐场。孩子们可以自由的奔跑，也可以玩滑梯、爬绳、睡吊床。游乐场被周边的树木保护起来，在夏日能尽享清凉。

水：主要设计元素

除了处在核心位置的哈尔堡之外，水在公园设计中也扮演着重要的角色。所有草地都被设计成简洁开放的空间，游客们可享受内港水域独一无二的景色。游客和居民可以在水边的台阶和码头上尽享阳光，或是休闲放松，观看水上船舶来来往往。由于上游港口的水并没有潮汐，游客们可以尽情坐在水边，十分安全。

乔木与灌木

环绕着公园，灌木花池和成群的榆树、赤杨树和柳树在公园和居住区之间形成了完美的过渡，将公私土地区域隔开。夏天，树木会投下一片清凉；秋天，树木能让整个公园变得金黄。水滨的灌木会不停的开花，在公园和水体之间形成了清晰的界限。对居民和游客来说，公园巧妙地结合了历史与现代，为小岛塑造了独特的形象。

SÜDPLATEAU
南方高地

Location: Berlin, Germany
Completion: 2014
Design: gruppe F Landschaftsarchitekten
Photography: gruppe F Landschaftsarchitekten
Area: 3,000sqm

项目地点：德国，柏林
完成时间：2014年
设计师：gruppe F景观建筑事务所
摄影：gruppe F景观建筑事务所
面积：3,000平方米

The "Südplateau" is located in the southern part of the Fritz-Schloß Park in Berlin Moabit. Despite its sunny and quiet location, it was only rarely used by local inhabitants. In order to strengthen the special amenity value of "the clearing on the hill", three lawn terraces are built. They are framed with granite blocks and serve as areas for seating and movement. A narrow paved pathway surrounds the plateau and is equipped with benches and rubbish bins. In addition, a large wooden deck and two deck chairs are placed on the lawn terraces. A new narrow climbing path is established as a sporty and playful short cut up the hill from the south-western part of the park. Boulders and high granite blocks invite to jump and climb and secure the steeper path sections.

In the meadow three grass terraces are stacked to be built around it a "Schlechtwetterweg" of small stone pavement, be placed on the new benches. In addition, a wooden platform and a few wooden chairs are planned that will invite you to stay. Because you previously only in the winter can look at the city from this place partially, individual sightlines are cut free. In the narrow northern area remains free of course, if at some point but still an observation tower is financed.

南方高地位于柏林弗里茨堡公园的南部。尽管这里阳光明媚，宁静宜人，当地居民仍很少踏足此地。为了进一步提升"山中空地"的独特休闲价值，设计师打造了三层草地平台。它们外围被花岗岩块包围起来，可以作为休息和活动区。一条狭窄的小路环绕着高地，路边配置了长椅和垃圾桶。此外，草地平台上还设置了大型木板平台和两个平台座椅。一条新开的上山小路是通往公园西南部的山丘的捷径，可供人锻炼。岩石和高大的花岗岩块吸引着人们上下攀爬，并且起到了陡坡保护作用。

三层草地平台叠加起来，一条由碎石铺设的小路环绕着平台，路旁还新设了长椅。另外，木板台和一些木座椅吸引着人们停下脚步，休息片刻。因为之前这里只有冬季才能俯瞰部分城市景象，设计师清空了一些视觉障碍。在狭窄的北面正规划建造一座观景台。

◀ 1. Plateau
2. Granite Blocks
3. Benches
4. Wooden Deck
5. Lawn Terraces
6. Wooden Deck Chairs
7. Climbing Path
8. High Granite Blocks
9. Boulders

1. 高地
2. 花岗岩块
3. 长凳
4. 木板平台
5. 草坪露台
6. 木板平台座椅
7. 攀登路
8. 高花岗岩块
9. 大岩石

CAMPA DE LOS INGLESES PARK
英人土地公园

Location: Bilbao, Spain
Completion: 2012
Design: Balmori Associates
Photography: Balmori Associates, Bilboa Ria, Borja Gomez Photography
Area: 300,000sqm

项目地点：西班牙，毕尔巴鄂
完成时间：2012年
设计师：Balmori建筑事务所
摄影：Balmori建筑事务所，Borja Gomez摄影公司
面积：300,000平方米

The park of the Campa de los Ingleses occupies the green space between Guggenheim Museum Bilbao and Deusto Bridge as well as the new buildings like the Library of the University of Deusto, the Auditorium of the UPV-EHU and the Iberdrola Tower, among others. This area, which in its time has been a British cemetery, the Athletic football ground and also a runway, is 25,000m² and was landscaped by Diana Balmori, who also designed the Plaza de Euskadi. The trees planted there (oak, holm oak and jacaranda) are also the most representative in Doña Casilda Park.

Campa de los Ingleses Park flows from the Guggenheim Bilbao Museum, unifying the Abandoibarra area of Bilbao and the Nervión River. The park's defining lines mark undulating paths that pull up to create a series of curving terraces. These topographic waves mediate a 10m elevation difference across the park. The terraces, ramps, stairs and walls flow into one another to sculpt a park that gracefully integrates the Mazarredo, Deusto Bridge, and the Plaza Euskadi with surrounding buildings and most importantly the Nervión River into a seamless urban experience.

Rather than using turf or "Industrial Lawn", a "Freedom Lawn" was planted by introducing various grass species, clover and wildflowers that fix nitrogen and reduce the need for pesticides. The paving designed by Balmori contains an additive called GeoSilex which absorbs CO_2; the paving was developed with the University of Granada and made entirely from industrial waste. The local newspaper referred to the park as "a new lung for the city."

英人土地公园占据着毕尔巴鄂古根海姆博物馆和德乌斯托桥之间的绿色空间，周围还环绕着德乌斯托大学博物馆、巴斯克大学的礼堂以及伊比德罗拉塔等建筑。这片区域曾经被作为英国人公墓、足球场以及跑道，总面积25,000平方米，由戴安娜·巴尔莫里进行景观设计。英人土地公园所栽种的树木（橡树、圣栎树、蓝花楹）也是多纳卡西尔达公园最具代表性的树木。

英人土地公园从毕尔巴鄂古根海姆博物馆开始延伸，连接了毕尔巴鄂的班多尔巴拿区和内维隆河。公园的边界线通过蜿蜒的走道向上形成了弧形阶台。这些波浪结构巧妙地化解了公园10米

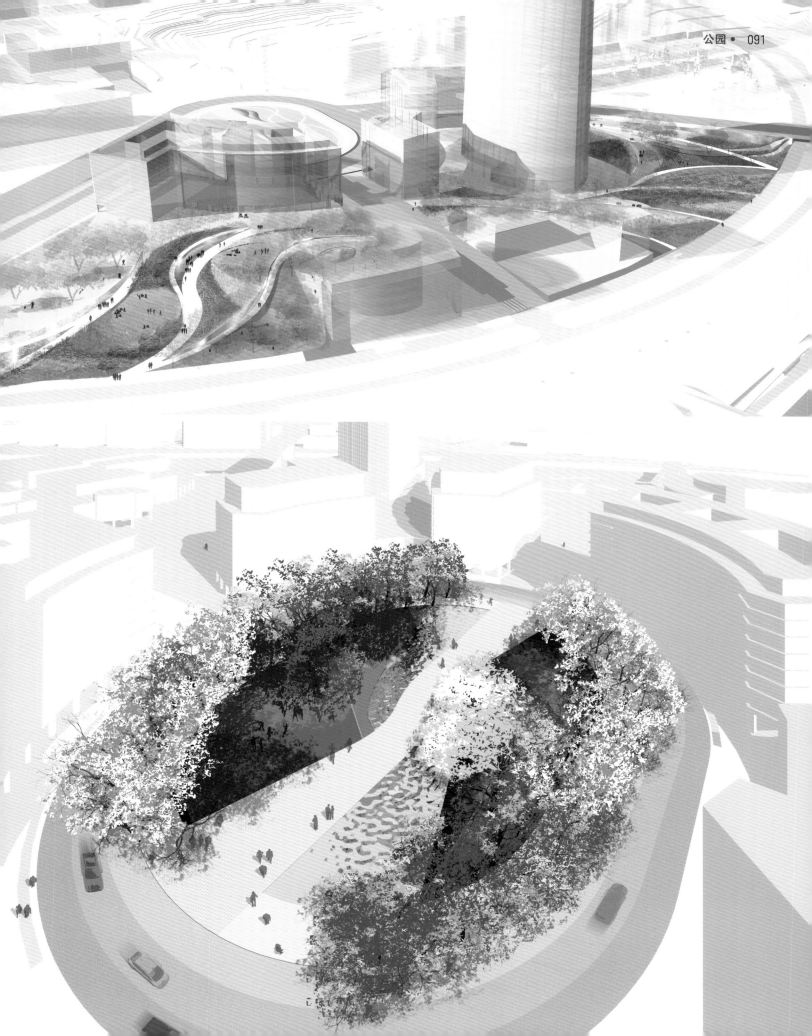

高的海拔差。阶台、坡道、台阶和围墙相互作用，共同塑造了公园，使德乌斯托桥、巴斯克广场与周边的建筑和内维隆河优雅地融合起来，形成了连续的城市体验。

设计师没有选择草皮或"工业草坪"，而是选择了"自由草坪"进行种植。他们引入了各种不同的草种、三叶草和野花，草坪植物的多样化能固定氮气，减少杀虫剂的使用。由Balmori建筑事务所设计的地面铺装包含一种名为GeoSilex的添加剂，可以吸收二氧化碳。铺装材料由设计师和格拉纳达大学共同开发，全部由工业废料制成。当地报纸将公园称为"城市的新肺"。

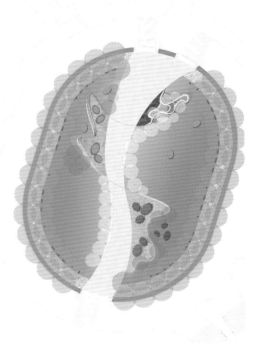

LEMVIG SKATEPARK
莱姆维滑板公园

Location: Lemvig, Denmark
Completion: 2013
Design: EFFEKT
Area: 2,200sqm

项目地点：丹麦，莱姆维
完成时间：2013年
设计师：EFFEKT景观事务所
面积：2,200平方米

In the spring of 2013, Lemvig Municipality faced a group of citizens eager to transform an empty industrial lot on the city's harbour front into an area of leisure and recreation. In order to meet the demands of the local population, EFFEKT worked closely with representatives from different user groups to develop a new type of urban space. The result of this collaboration was an integrated skatepark + urban park that offered a range of programmatic features and recreational opportunities. Set in beautiful surroundings, the park has created a new social space in Lemvig, attracting skaters and families from the entire region.

"The harbour, having displaced most of its activity along the coast, had become a residual wastescape of maritime activity. By envisioning the Skate+Park as a social gathering space that would attract people of all ages and interests, we believed it could become a catalyst for revitalisation that would re-brand the harbour front as a recreational hub and re-introduce the harbour as an important asset to the city." says Mikkel Bøgh of EFFEKT.

"From the start, the project would need an array of ingredients to differentiate itself from the grey, black and rust-tinted surfaces of the immediate surroundings – the consequence of a downturn in the local fishing industry. By challenging the typology of the skatepark – an otherwise mono-functional greyscape – alongside a thorough investigation on the dichotomy of surface in public space, the architects were able to design a hybrid platform that would accommodate a multitude of social and recreational activities. Skateboarding originated in streets, co-existing with a multitude of other urban activities. As it grew in popularity and commercialised, the sport and culture was moved into these grey parks were it became isolated from the same city that originally fuelled, challenged and inspired the skaters. By merging skateboarding with a multitude of other recreational activities and re-introducing the (skate) culture back into the heart city centre, we feel the both skateboarders and other groups of the population will benefit greatly from this new co-existance on the harbour – and potentially breathe new life into an otherwise abandoned area with great potentials."

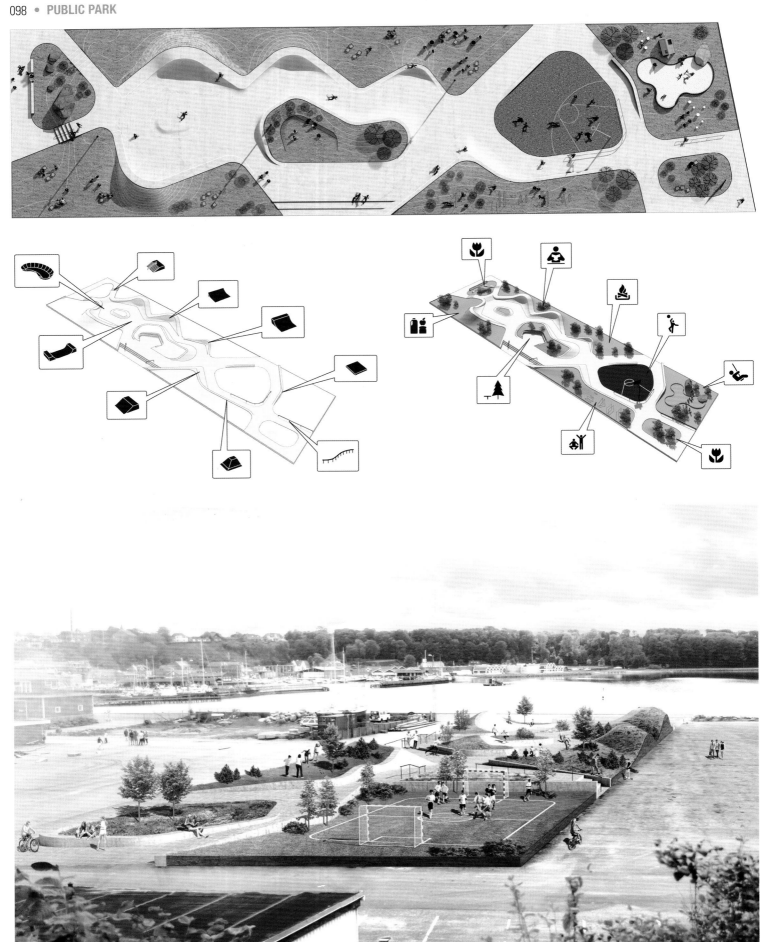

2013年春，莱姆维市政府面临着市民想要将港口上一处工业空地改造成休闲娱乐区的热切渴望。为了满足当地居民的需求，EFFEKT景观事务所与来自各个用户群体的代表进行了紧密合作，共同打造了一个新型城市空间。滑板公园与城市公园的组合为市民提供了一系列功能设施和休闲选择。在优美的环境中，公园为莱姆维市营造了全新的社交空间，吸引着整个地区的滑板爱好者和家庭游客。

EFFEKT 景观事务所的米克尔·波尔称："港口已经丧失了自身的价值，成为了一个废弃的海洋景观地块。滑板公园的设计能打造一个吸引各个年龄段人群的社交空间，我们坚信它将成为复兴港口为娱乐中心的催化剂，使其重新融入城市生活。"

"项目迫切地需要一些元素来使其与灰黑色、锈迹斑斑的周边环境（本地渔业衰退的结果）区分开。为了打造一个成功的滑板公园，设计师不仅在公园中加入了滑板坡道等灰色景观元素，还设计了一个混合平台，可以进行各种各样的社交娱乐活动。滑板运动起源于街头，与多样化的城市活动并存。随着滑板的流行化和商业化，滑板运动和文化逐渐搬进了专门的滑板公园，使滑板者与城市生活脱离开。项目实现了滑板运动与多样化娱乐活动的融合，并且将滑板文化重新引入了城市中心。我们认为滑板爱好者和其他城市群体都会从港口的新形态中获益，从而进一步为这片曾经被废弃的区域带来生机。"

THE PULSE PARK
脉动公园

Location: Skanderborg, Denmark
Completion: 2012
Design: CEBRA
Photography: Mikkel Frost
Area: 2,235sqm
项目地点：丹麦，斯堪德堡
完成时间：2012年
设计师：CEBRA景观事务所
摄影：米克尔·弗罗斯特
面积：2,235平方米

Kildebjerg Ry is a very popular residential area for families mainly because of the beautiful surrounding countryside that lends itself to a wide variety of outdoor activities. Since the community wanted to expand this attractive system of leisure and sports facilities, CEBRA was asked to design three publicly accessible and innovative activity zones – each aiming at a different type of activity.

The aim of the project was to create optimal conditions for physical activity and play, which form an integral part of the landscape, the area's additional leisure activities and the residential area itself.

The Pulse Zone is a literal bulge on the existing paths as it prompts horizontal movement and primarily addresses runners, skaters and mountain bikers, who use the network of paths and tracks around Kildebjerg Ry for exercise. The Pulse Zone's asphalt track flows into the zone's centre area in the form of hills, bulges and bowls that incite to obstacle races and artistic exercises.

The Play Zone invites to both play and physical exercise in a forest of different functions that relate to organised as well as freestyle activities. This forest is made up from three concentric circles. The centre is primarily for play, climbing and exercise in large tree-like elements. The intermediate circle is for outdoor fitness and offers a series of primitive versions of well-known training devices. The outermost circle consists of a recreational zone for social activities and gives room for meeting, relaxing, picnicking etc. close to a fire place.

The third zone complements these active zones and provides a more contemplative space: The Zen Zone is all about relaxation and activating all your senses. The zone provides a green and quiet environment for activities like yoga, Pilates and meditation, partly shielded under trellises. It is placed on an artificial island in a small lake, which is surrounded by a garden that stimulates and changes the sensory perception as you move through the zone.

104 • **PUBLIC PARK**

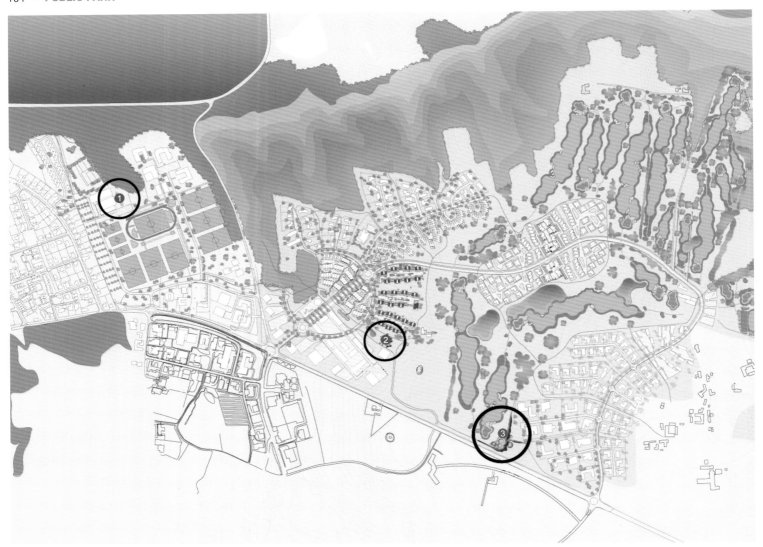

▲ 1. Play Zone
   2. Zen Zone
   3. Pulse Zone

   1. 游戏区
   2. 禅意区
   3. 脉动区

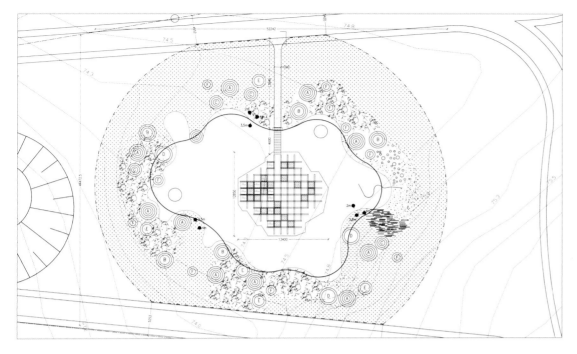

吉尔德比山社区是一个深受欢迎的住宅区,以家庭住户为主,享有优美的乡村景色和丰富多样的户外活动空间。社区希望对休闲运动设施进行扩建,因此委托CEBRA景观事务所设计了三个创新型公共活动区,每个活动区对应一种不同类型的活动。

项目的目标是为体育活动和游戏提供良好的条件,并使其融入该地区的景观、其他休闲活动以及住宅区。

脉动区高出原有地面,主要针对跑步者、滑冰者以及山地车骑手,他们利用环绕社区的道路和跑道网络进行锻炼。脉动区的柏油跑道以小山的形式流入中央区域,形成凹凸不平的运动竞技空间。

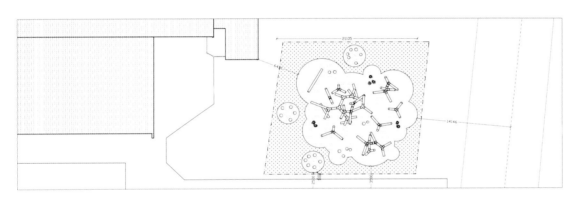

游戏区设有一系列不同的功能设施,可供人们有组织地锻炼或自由活动。器材森林由三个同心圆组成。圆心可在大树造型的设施上游戏、攀爬和锻炼。中间是户外健身区,提供了简单的训练设施。外圆由休闲区构成,可以进行聚会、休闲、野餐等社交活动,靠近火炉。

与前两个区域相比,第三个区域更偏向于沉静,是能够让所有感官得到放松和活跃的禅意区。这里为瑜伽、普拉提、冥想等活动提供了绿色安宁的环境,被保护在格架结构之下。它被设置在一个小型湖中央的人工岛上,被花园所环绕,能够激发人的感官知觉。

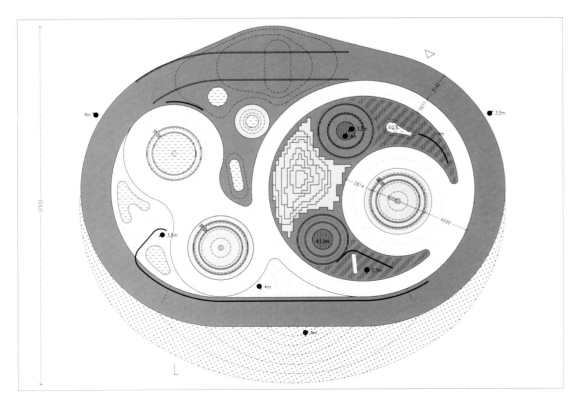

Location: Rotterdam, the Netherlands
Completion: 2011
Design: OKRA landschapsarchitecten bv
Photography: OKRA landschapsarchitecten bv
Area: 53,000sqm
项目地点：荷兰，鹿特丹
完成时间：2011年
设计师：OKRA景观建筑设计私人有限公司
摄影：OKRA景观建筑设计私人有限公司
面积：53,000平方米

ROTTERDAM WESTERKADE AND PARKKADE 鹿特丹韦斯特码头和公园码头

Rotterdam City on the Maas

The vision for the city of Rotterdam is that it should be again the city engaging with the river Maas. The docks play an important role as a transitional area and recreational support. Recreation strengthens the relationship between city and water. In particular, the northern quays have been given a green and softer atmosphere, which reinforces the long line of the series of quays.

The Westerkade and Parkkade are part of the northern shore and form an almost unbroken line along the Rotterdam waterfront. Unity in design makes both docks into a continuous whole, without diminishing the different characters and ways of use. The Parkkade still serves as a landing dock for cargo. It is one of the few places in the centre where the atmosphere as a world port of Rotterdam is still recognised. Currently, the park is still cut off from the Maas; this is where the vision indicates how the relationship with the lower quay can be strengthened.

Westerkade

The Westerkade has become a place to stay, with an attractive programme. The idea of is based on functional flexibility. An inviting place, which provides both intimately scaled and large scale spaces and has an ability of adapting to larger and smaller events. To achieve this, the quay is cleared of parking creating new opportunities for Westerkade to become an attractive place to stay and a green atmosphere.

The new layout provides a basic profile with a clear beginning and ending with its own zoning. The existing trees are retained wherever possible with a comprehensive single row of trees on the side of the road. At head of the ferry terminal there is a new pavilion with a terrace overlooking the river Maas. The existing harbour for the water taxi is transformed into an attractive place along the Maas, with a wide staircase to the water and a wooden platform as a new stop for the water taxi. Between the edges of the docks there are a series of smaller green spaces created with beds of ornamental grasses and perennials. In these places, both public sitting areas and terraces can be found.

城市公共空间 • 109

On the quay elm trees determine the image as seen from the water and from the south of the Maas. In the existing double row of trees some trees are missing. These will be supplemented where necessary by new elms. The green image of the pontoon, not only created by trees, but also by a green ground. This surface consists of green grasses supplemented by perennials and bulbs. The grasses have a striking appearance and beautiful flowering spikes waving in the wind, while the perennials to provide a more varied picture throughout the seasons.

Sustainability

For Westerkade and Parkkade the existing natural stones have been recycled. In the concept only the walking zone along the water and the two heads are made form new material. For the rest of the pavement on the quay the existing stone is reused.

马斯河上的鹿特丹港市

鹿特丹的城市景观应该进一步与马斯河建立起联系。作为城市与运河之间的过渡区域、同时又为人们的娱乐活动提供支持的码头，在这种联系建立的过程中有着重要的作用。人们的娱乐活动加强了城市与河水之间的关系。特别值得注意的是，设计师为北部的码头营造出绿色柔和的环境，从而使一长串的码头序列得到了强调。

作为北部海滨的一部分，韦斯特码头和公园码头形成了一条几乎贯通整个鹿特丹沿岸的滨水景观线。统一的设计将两个码头连接成为一个整体，却没有削弱它们各自不同的特征和用途。公园码头仍然作为货物的登陆码头。它也是少数位于中央的空间之一，在这里，鹿特丹作为世界港口的特征仍旧清晰可见。目前，公园与马斯河仍未连通；这表明公园与下游码头之间的关系有进一步加强的空间。

韦斯特码头

韦斯特码头在设计师的改造下成为一个停留空间，其中布置有吸引人的元素。设计的理念基于功能的灵活性。码头作为一个邀请空间，需提供大小两种不同尺度的空间，并可以举行大型或小型的活动。为实现这一目标，设计师移除了码头上的停车场，这一举措为打造韦斯特码头成为一个吸引人停留的空间，营造绿色的氛围创造了新的机遇。新的布局明确了公园的基本轮廓，以及公园起点与终点的区域划分。现存的树木尽可能被保留下来，组成单排的混植行道树。在渡轮码头的前端，设计师添加了

一个带平台的亭子，游客可以在亭子上眺望马斯河全景。原有的水上出租车停泊港被改造成马斯河沿岸一个非常吸引人的场所，这里有一条通向水面的宽大台阶，还有一个木制的平台作为水上出租车新的停靠点。在码头的边缘之间，有一连串面积较小的绿地，绿地中栽植着观赏草和多年生植物。在这些地方，都可以找到公共座椅和平台。从水面上或者马斯河南部眺望码头，看到的是满眼的榆树。现存的双排树木中有一些位置有空缺。设计师将在空缺的位置上补植新的榆树。浮动码头的绿色景观，不只由树木创造，也由绿色的草坪创造。设计师在原有的绿草地中补植了多年生植物和鳞茎植物。草坪景观十分引人注目，漂亮的花穗在风中摇曳，多年生植物使这里一年四季的景象更加丰富。

可持续性

韦斯特码头和公园码头上原有的天然石材被再次利用。根据设计师的构想，只有沿着水岸的步行区和公园的两个端点的地面使用新材料铺设。码头上其余的人行道都重复利用原有的石材。

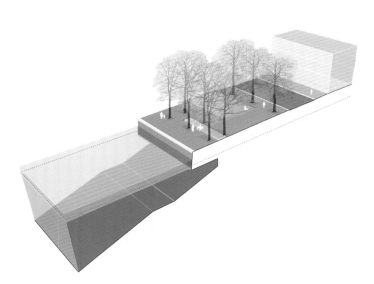

BLOKHOEVE
布罗克霍温

Location: Nieuwegein, the Netherlands
Completion: 2012
Design of skate and play: Carve
Park design: Dijk&co landscape architecture
Photography: Carve

项目地点：荷兰，尼沃海恩
完成时间：2012年
滑冰道和游乐设施设计：雕刻景观设计事务所
公园设计：Dijk&co景观建筑事务所
摄影：雕刻景观设计事务所

The Park 'island-West' is the community park of the new district Blokhoeve. It is a hilly grassland with loosely scattered trees. Although it is a new park, the trees are quite old; several years ago the existing hungarian oak and linden trees were replanted, stored and placed back in the park after the construction period. The pathsystem is designed as a 'liquid gel' that connects all entrances and includes the sports- and playgrounds as a logical but playful pattern. The old location contained an old running track that was integrated in the new design. It now is a fresh running and skating track, that encloses the new sports and play functions.

The skate and play landscape was designed by Carve. The skate objects 'stick' to the inside edge of the track, and are made of light coloured concrete that forms a nice contrast with the dark asphalt of the track. Being a neighbourhood park with a sporty character, the choice was made to integrate game and sport in the play objects, too. The play object, with its vertical tree trunk forest, balances the vertical direction of the mature trees in the park. The play cubes, which hang tilted between trunk forest, are designed from the scale and perception of the child. The exterior of the same play objects, however, is a professional boulder wall, so parents are challenged to climb on it. The playcubes are placed quite high to prevent small children to climb on the outside. Nevertheless, children can crawl and climb from one cube to the other through elevated climbing paths. Two worlds are united in one object: they don't blend but they do meet. With this, the children's playground also becomes a playground for adults.

"海岛西园"隶属于布罗克霍温新区，是一座社区公园。公园的主体是一个坡地草坪，草坪各处松散地生长着几棵大树。虽然这个公园是新近建成的，但是其中的树木却非常古老：几年前，施工人员将基地上原有的匈牙利橡树和椴树移走，并保护起来，在建设工期结束后，又将它们移植回了公园。公园道路系统的形态如同"液体凝胶"，将所有的入口、运动场和游乐场连接起来，形成了一个逻辑清晰且妙趣横生的平面图案。在基地原址上，有一条老旧的跑道，设计师将其整合进新公园的设计中。现在它是一条焕然一新的跑道，并加入了新的运动和娱乐功能——也可以当作滑冰道使用。

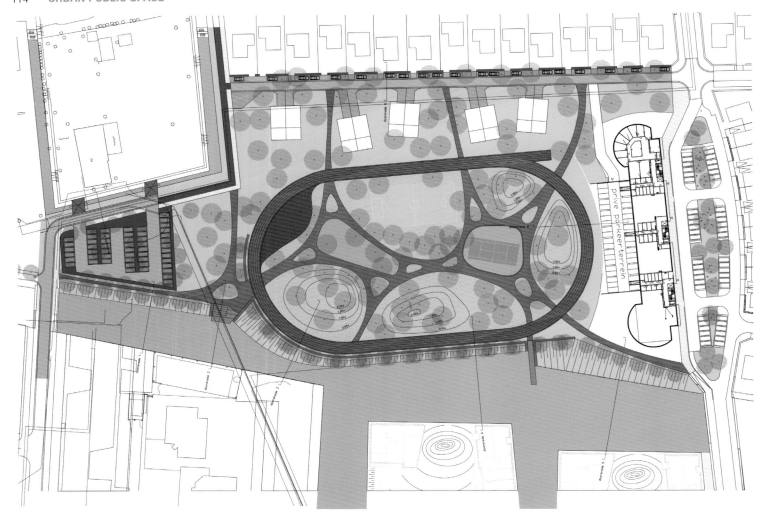

公园的滑冰道和游乐设施景观由雕刻景观设计事务所负责设计。滑冰道"粘"在跑道的内边缘，它的混凝土外观呈现出的浅色调，与跑道的沥青黑形成鲜明的对比。为了突出这个社区公园的运动特性，设计师将游戏与运动功能融入游乐设施的设计之中。这个游乐设施，由竖向的"树干森林"组成，"树干"的方向与园中那些成熟树木的方向保持一致，形成了和谐统一的景观。在"树干森林"之间，悬挂着许多倾斜的游戏立方体，它们依照孩子的身高和感官而设计。然而在游乐设施的外立面，却是专业的大卵石墙，家长们可以在墙面上攀岩。为了防止小孩子从外面爬上这些立方体，设计师将它们的位置设计得非常高。但是，孩子们可以在立方体内部通过悬空的隧道匍匐前进，从一个立方体，爬至另一个立方体。成人和儿童的两个世界，通过这个立方体连接：它们相接却不混合。通过这种形式，这个儿童游乐场也成为了可供成人运动的场所。

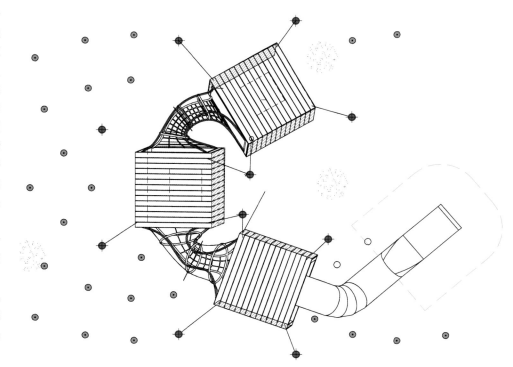

城市公共空间 • 115

KAVEL K
多K青年溜冰运动场

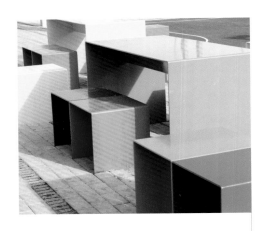

Location: The Hague, the Netherlands
Completion: 2014
Design: Carve
Photography: Marleen Beek
Area: 1,650sqm

项目地点：荷兰，海牙
完成时间：2014年
设计师：雕刻景观设计事务所
摄影：玛伦·比克
面积：1,650平方米

"Kavel K" is situated on a triangular plot, boxed in by a railwaytrack and a connecting road. It is a skating, sports- and youth facility which attracts a wide range of user groups. The public space and the building are designed as a unity; the facade and the skate-cradle even "melt together".

Kavel K is one of the three skate facilities that were originally planned in the urban layout plans for Leidschenveen-Ypenburg, which started in the 90s. The location is a typical surplus space; the tapered terrain is wedged between a railway track and a connecting road, at the edge of one of the largest Vinex-neighbourhoods in the Netherlands. Previously, only skating was planned here, but because of demographic changes the need for a youth centre grew; the young families that moved here now have adolescent children.

Carve was asked to design both the amenities and the building. They thought it to be of great importance to create a building that presents itself as one with its surroundings, both visually and functionally. But how can skating, sports and a youth centre designed in such a way as to create a whole?

The small strip of land is divided into three zones: skating, youth centre and sports. By positioning the youth centre in the middle of the zone, a front- and backside are created, between which the building forms the hub. The entry zone is flanked by the skate facility, which is elevated half a meter above ground. By raising the skating area, a sitting edge is created along the entry zone. Furthermore, by raising the skating, the entrance can be reached without being hindered by skaters. As a contrast to the front, the multifunctional sportscourt in the back is sunken and its edges can be used for activities.

The facade and the integration of the skating facility are an essential part of the design. The cradle – an eyecatcher at the front side – is integrated into the facade. By doing so, the facade and skatepool become one entity.

冰场地的标高,赋予了入口两侧的冰场边缘可坐与休息的功能。而且,得益于这样的设计方式,人们出入活动中心时可以避开滑冰者的干扰。不同于前方的滑冰场地,后方的多功能运动场是下沉式的,它的边缘可以用于活动。

建筑立面和滑冰设施的融合是设计的关键部分。摇篮状溜冰池——作为引人注目的正面景观——与建筑立面结合在一起。通过这种方式,建筑立面和溜冰池成为一个实体。

第二条设计原则是,设计师意识到,建筑的立面不可避免地会布满孩子们的涂鸦。但是他并没有将这视为一个麻烦,而是在设计中预料到了这点并加以利用。建筑的立面由大面积的混凝土元件组成,混凝土元件上面压制出"盲文图案"。墙上的涂鸦可以被抹去,但是在盲文图案凹陷的圆圈中还会留下涂鸦的痕迹。因此,立面成为一个画布,上面不断变化的颜色与图案,记录着建筑的历史。

第三条设计原则是建筑使用上的灵活性,特别是在平面的设计上,这种灵活性尤为突出。室内空间的设计保持着粗犷简单的风格,设计师为建筑未来的使用者留出了足够的发挥空间。景观核心和地面均使用对比鲜明的颜色,墙体内部衬有耐用的衬垫材料面板。核心景观周围巨大的滑动门,为人们通过多种方式自由分割空间创造了可能。此外,建筑物面向滑冰场一侧和运动场一侧,各有一个入口。现在只有一个入口开放,但是在将来两个入口都将投入使用。这样,不同的使用群体——无论是滑冰者、爱冒险的青年、喜爱运动的青少年,还是前来上打字课的摩洛哥母亲们——都可以不受其他人的干扰,各自独立使用相关的场地和设施。

LAAN VAN SPARTAAN
斯巴达大道

Location: Amsterdam, the Netherlands
Completion: 2013
Design: Carve
Photography: Marleen Beek
Area: 437sqm
项目地点：荷兰，阿姆斯特丹
完成时间：2013年
设计师：雕刻景观设计事务所
摄影：玛伦·比克
面积：437平方米

Laan van Spartaan is a new city development just outside the ringroad of Amsterdam. On the former site of footballclub VVA Spartaan and Sportshall Jan van Galen more than 1,000 housing units will be built until 2016. Renowned architects – among which Claus en Kaan, Dick van Gameren, Diederen Dirrix, MVRDV and DP6 – designed the housing blocks, a new sportshall, offices, a school and community facilities. The dike along Laan van Spartaan, on the Willem Augustinstreet, was designated as playzone. The 200 metre long strip, however, is squeezed in by a bike path and a lowered road, which is why the location actually could not contain any play equipment broader than 1.50 m.

The limitations of the location were taken as a starting point for the design. By raising the play objects, increasing the height and giving them a minimal footprint, the maximum of the limited space was used. Three playzones, each 25 metres long, offer playoptions for different age groups. The northern space is designed especially for younger children, while the middle zone – with play houses and hammocks – attracts somewhat older children. The southern and most challenging playzone contains a climbing garland, which curls up to more than 5 metres height, offering a panoramic view on the whole strip. Only daredevils will go up here! Whilst the initial concept of the playstrip was there at the very beginning – and didn't change much during the process – the technical engineering turned out to be a great challenge, due to all different angles, heights and cantilevers.

Children from the neighbourhood, who are organised under the header "Portiekportiers" (meaning as much as "porch-porter"), have been pro-active when it comes to organising activities in their community. During the design process they were consulted several times and during the brainstormsessions they chose the name "Willem Augustinpark" for the playzone. A community that does not have roots, yet, greatly benefits from spaces were children can play and parents can meet. Laan van Spartaan fulfills this role with fervour. Furthermore, the visibility of the climbing objects from afar contributes to the fact that also children from surrounding neighbourhoods discovered this playspace.

斯巴达大道是紧邻阿姆斯特丹环城公路外围的一个新建成的城市居住区。到2016年为止，在基地的原址、VVA斯巴达足球俱乐部和扬·盖伦体育馆上，将新建1000座住宅单元。著名的建筑师和建筑事务所——包括克劳斯·卡恩、迪克·范·哈默伦、Diederen Dirrix事务所、MVRDV事务所和DP6事务所——参与了住宅楼、一座全新的体育馆、写字楼、一所学校和社区设施的设计。沿着斯巴达大道的路堤，在威廉·奥古斯丁街，设计师规划出一块游乐区。但是，这块200米长的条形地带，被一条自行车道和一条下沉的道路占据得所剩无几，正是由于场地太过狭窄，这个地块不能容纳任何宽度大于1.5米的游乐设施。

设计师将基地的局限性作为设计的出发点。他们抬高了游乐设施，增加了它们的高度并使它们的着地面积最小，通过这样的设计，有限的场地得到了最大化的利用。三个游乐区，每个长度为25米，为不同年龄组的群体提供了多样的游戏选择。北边的场地特别为幼儿设计，中间的区域——其中有游戏房和吊床——吸引了稍微大一点儿的孩子。南部是最具挑战性的游乐区，它包含了一个攀岩花环，花环螺旋状上升，高度超过五米，站在花环上面，可以俯瞰整个条形地带的全景。只有勇敢的人才能到达那里！同时游戏带的最初概念也是起源于此——在设计过程中，这一概念没有发生太大的变化——技术工程反而成为设计师面临的巨大的挑战，因为游乐设施有着不同的角度、高度和悬挑，结构复杂多变。

当社区举办活动时，附近的孩子们积极地参与其中。在设计过程中，设计师也多次询问了孩子们的意见，在头脑风暴会议上，孩子们为游乐区取名叫做"威廉·奥古斯汀公园"。这是一个没有根基的社区，但是因为有了这个游乐空间，孩子们可以在这里玩耍，家长们可以在这里会面。斯巴达大道带着热情，完成了它的使命。更重要的是，这些攀爬游乐设施从远处就看得见，周围居住区的孩子们很快也发现了这块游乐场地。

城市公共空间 • 125

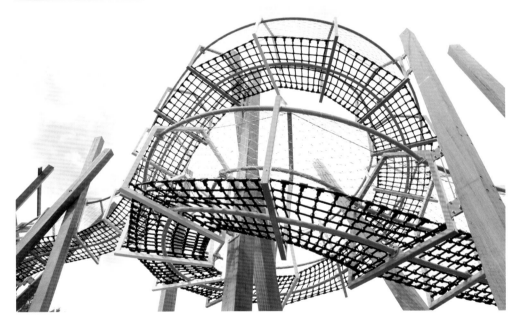
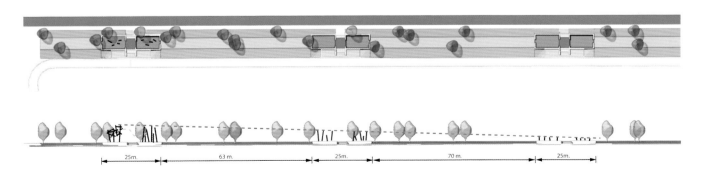
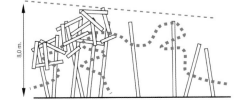

WATER SQUARE BENTHEMPLEIN
本瑟姆水广场

Location: Rotterdam, the Netherlands
Completion: 2013
Design: DE URBANISTEN
Photography: DE URBANISTEN, Ossip van Duivenbode, Jeroen Musch
Area: 9,500sqm

项目地点：荷兰，鹿特丹港市
完成时间：2013年
设计师：德·瑞班斯滕设计公司
摄影：德·瑞班斯滕设计公司，奥斯皮·范·德文布登，吉荣·穆什
面积：9,500平方米

The water square combines water storage with the improvement of the quality of urban public space. The water square can be understood as a twofold strategy. It makes money invested in water storage facilities visible and enjoyable. It also generates opportunities to create environmental quality and identity to central spaces in neighbourhoods. Most of the time the water square will be dry and in use as a recreational space.

On the Benthemsquare the first water square has been realised. In an intense participatory trajectory with the local community we jointly conceived ideas about the square: students and teachers of the Zadkine college and the Graphic Lyceum; members of the adjacent church, youth theatre and David Lloyd gym; inhabitants of the Agniese neighbourhood, all took part. In three workshops we discussed possible uses, desired atmospheres and how the storm water can influence the square. All agreed: the water square should be a dynamic place for young people, lots of space for play and lingering, but also nice, green intimate places. And what about the water? This had to be excitingly visible while running over the square: detours obligatory! The enthusiasm of the participants helped us to make a very positive design.

Three basins collect rain water: two undeep basins for the immediate surroundings will receive water whenever it rains, one deeper basin receives water only when it consistently keeps raining. Here the water is collected from the larger area around the square. Rainwater is transported via large stainless steel gutters into the basins. The gutters are special features, they are oversized steel elements fit for skaters. Two other special features bring storm water on to the square: a water wall and a rain well. Both dramatically gush the rain water visibly onto the square. The rain well is designed as a special beginning to the stainless steel gutter lifting itself from the ground. This well brings the water from the adjacent building into the gutter. The water wall brings the water from further away into the deep basin. Here a rhythm of waterfalls is being directed in relation to the amount of water falling from the sky. Two more water extras complete the picture. An open air baptistery is placed next

128 • URBAN PUBLIC SPACE

to the church that is situated on the square. Here a small fountain starts from which the water meanders over the square into one of the undeep basins. And in the deep basin we "join the pipe" and plant a drinking fountain for all thirsty athletes to enjoy.

After the rain, the water of the two undeep basins flows into an underground infiltration device and from here gradually seeps back into ground water. Thereby the ground water balance is kept at level and can also cope with dry periods. This helps to keep the city trees and plants in good condition which helps to reduce urban heat island effect. The water of the deep basin flows back into the open water system of the city after a maximum of 36 hours to ensure public health. All the storm water that has been buffered does not flow into the mixed sewage system anymore. Like this the conventional mixed sewage system is relieved and lowers the frequency of his relatively dirty water to overflow in the open water whenever it reaches its buffering capacity. By separating storm water gradually from the black water system with each intervention, the entire system step by step moves towards an improvement of the overall quality of the open water in the city.

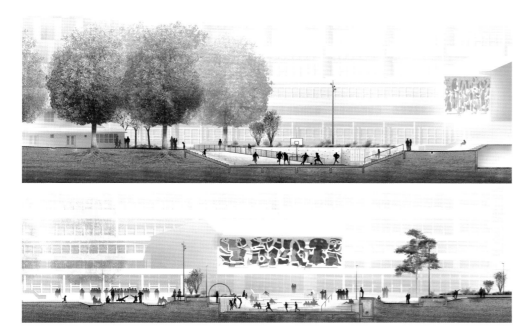

本瑟姆水广场结合了储水和提高城市公共空间水质两方面功能。水广场可以被理解为是设计师的双重设计策略。它使投资于储水设施的资金实现可视化，且产生了宜人的效果。同时也为提升社区中央空间的环境品质和个性特征创造了机会。大部分时间里，水广场的地面保持干燥，人们仅把它作为一个娱乐空间来使用。在本瑟姆广场，设计师首次实现了水广场的构想。设计过程中，当地社区居民热情参与其中，与设计师们共同构思了关于广场设计的想法：扎德金大学以及平面艺术学会的学生和老师们；临近教会的成员，青少年和大卫·劳埃德体育馆；阿格涅色社区的居民，都参与了研讨。他们在三间工作室中，探讨了广场可能的使用方法，渴望创造的氛围，以及雨水会如何影响广场的使用。最终所有人一致同意：水广

场应该成为一个动态的空间，为年轻人提供大量游戏和漫步的场地，但是同时它也需要是一个友好、绿色、亲密的空间。那么如何处理水呢？答案是：水流经过广场时，要产生令人兴奋且引人注目的效果：无疑我们需要曲折迂回的水流！参与者的热情帮助设计师们达成了非常积极的设计。

设计师设计了三个用来收集雨水的水池：每当下雨时，两个不深的水池都将收集周围的雨水，而另一个更深一点的水池，只在持续下雨时才收集雨水。深水池从广场周围收集雨水的范围比浅水池更大。雨水通过大的不锈钢水槽流入水池中。水槽特征鲜明，它们如同加大号的钢制滑冰道。另外还有两个将雨水引入广场

的特殊景观元素：一座水墙和一口雨水井。雨水从这两个水景小品中涌出，落到广场上，效果十分引人注目。雨水井作为不锈钢水槽特殊的起始点，突出于地面。这口井从邻近建筑物中将水导入水槽。水墙则从更远的地方把水引到深水池。水墙上的瀑布，随着天空降雨量的不同变化着节奏。还有两个额外的水景设施共

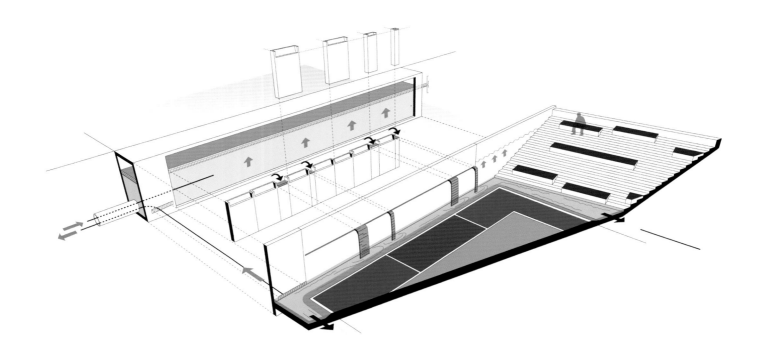

同构成了这幅景观。紧挨着广场上教堂的,是一个露天洗礼堂。这里有一个小的喷泉,水从喷泉中涌出,在广场上蜿蜒流淌,最终流入一个较浅的水池。在深水水池里,设计师们"将管道连接了起来",并安置了自动饮水器,供所有口渴的运动员使用。

下雨后,两个较浅的水池中的水流进地下渗滤设施,再从那里逐渐渗透回地下,成为地下水。通过这种方法,地下水在一定的水平上保持着平衡,同时也能应对干旱期的缺水情况。这有助于保证城市中树木和其他植物保持良好状态,从而减轻城市热岛效应。深水池中的水最多在36小时之后便流回到城市的开阔水面系统,以保证良好的公共卫生状态。所有这些得到缓冲的雨水,将不再流回混合污水处理系统。通过这样的设计,减轻了传统的混合污水处理系统负担,也相对降低了当系统达到缓冲容量临界值时,脏水溢出到开阔水面的几率。通过各项干涉措施,设计师成功将雨水从废水系统中逐步分离,使得城市开阔水面整体水质一步一步得到了提高。

STATION AREA VOORBURG
福尔堡站区景观设计

Location: Leidschendam-Voorburg, the Netherlands
Completion: 2013
Design: Posad
Photography: Christian van der Kooy / Bart van Hoek
Area: 34,000sqm

项目地点：荷兰，莱德斯亨丹－福尔堡
完成时间：2013年
设计师：帕萨德景观设计事务所
摄影：克里斯蒂安·范·戈伊 / 巴特·范·赫克
面积：34,000平方米

Over the last four years the station area of Voorburg has been transformed. The highway between Utrecht and The Hague was constructed in the eighties and over-built a large part of the area. The structure and organisation of the station area was very unclear and messy. The new design of the station area of Voorburg connects the station with the historical centre of Voorburg and the area of the Binckhorst, that is being redeveloped. The plan consists of different elements: the station square, 'the Strip', bus station, P+R, historical garden, pedestrian bridges, a park and a tram stop with a waiting space.

The station area of Leidschendam-Voorburg is a complex area, largely located underneath the infrastructure of the highway A12 and the railway tracks, and on top of a seventeenth century estate of Constantijn Huijgens: Hofwijck.

The main design task was to establish a balance between the historical layers of the site and to connect the areas around the site. In the design Posad combined the oldest layer of the seventeenth century garden (following the geometry of Vitruvius) and the infrastructure that was built on top of this garden. Architect Carel Weeber elevated the highway and the railway in the eighties and he restored the sightlines in the area, without taking the diagonal direction of the infrastructure into account. Posads design re-establishes the balance between the contrasting structures and strengthens both.

The transit oriented design consists of different areas with their own qualities. The square, bus station, P+R and the park, are all connected by 'the Strip'; a spacious pedestrian path that connects the historical centre of Voorburg with the business park the Binckhorst in The Hague. The Strip helps the traveller to find his way in a natural and logical manner, when he gets out of the station. The light pavement counterweights the large and heavy viaduct and the Strip has a certain rhythm with trees, lighting elements and benches. Parking places that used to block all routes from the station are moved and a lot of space is cleared for pedestrians and bicyclists.

城市公共空间 · 133

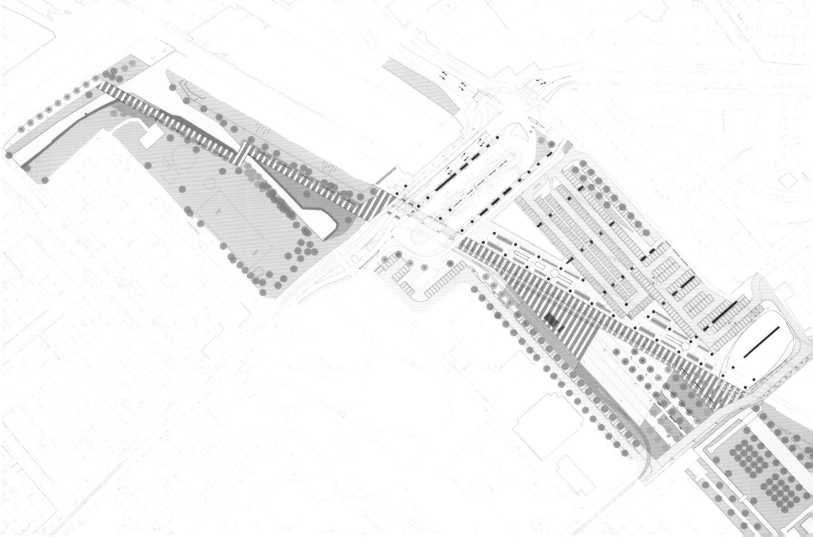

在过去四年间，福尔堡车站区域经历了改建。一条修建于20世纪80年代、连接着屋特勒支和海牙之间的高速公路，占据了这一区域的很大一部分。火车站区的结构和组织非常混乱不清。福尔堡车站区域新的设计将车站与福尔堡城的历史中心以及正在重建的宾克霍斯特地区连接了起来。设计包含了不同的元素：车站广场、"主干道"，公共汽车站，P+R换乘站，历史花园，步行天桥，一座公园和一座带有等候空间的电车站。莱德斯亨丹－福尔堡站区是一个综合区域，它的大部分位于A12公路基础结构和铁路轨道的下方、一座十七世纪康斯坦丁·惠更斯的地产——霍夫维克的顶端。

设计师的主要设计任务是建立车站与基地的历史层次之间的平衡，以及车站与周围区域的连接。在设计中，帕萨德景观设计事务所将最古老的层次——一座十七世纪花园（依从维特鲁威的几何结构）和建造在这座花园之上的基础设施结合了起来。建筑师卡诺·韦伯抬高了八十年代修建的公路和铁路，并还原了这一区域除基础设施的对角线方向外的视线。帕萨德事务所的设计重建了这两个结构差异明显的元素之间的平衡，并对两者都起了强调作用。交通导向设计的范围囊括了不同区域，并且各个区域具有各自的特点。广场、公共汽车站、P+R换乘站和公园，所有这些元素都通过"主干道"连接；一条宽阔的人行通道将福尔堡的历史中心和海牙宾克霍斯特的商业园区连接起来。当旅行者走出火车站时，主干道通过自然和富有逻辑的方式帮助他们找到方向。轻质的人行道平衡了大而沉重的高架桥，主干道上的树木、照明设施和长椅以一定的节奏重复布置。设计师将阻挡通向火车站道路的停车空间移除，为行人和自行车创造了许多空间。

THE RAVELIJN BRIDGE
拉维利因堡垒岛之桥

Location: Middelburg, the Netherlands
Completion: 2014
Design: RO&AD Architecten
Photography: RO&AD Architecten

项目地点：荷兰，米德尔堡
完成时间：2014年
设计师：RO&AD建筑设计事务所
摄影：RO&AD建筑设计事务所

The Ravelijn "Op den Zoom" is a fortress-island of the city of Bergen op Zoom in The Netherlands which is made in the beginning of the 18th century by Menno van Coehoorn, a famous fortress builder. This is the only "ravelijn" of him still present. The fortress was originally only accessible by boat , so supplies and soldiers had to be rowed the 80 metres to the fortress. The original entrance is still present just above the waterline At the end of the 19th century the fortress lost its defensive function. In 1930 a raised wooden bridge was added. Nowadays the island-fortress is mainly used for small public and private events. The assignment was to make a second pedestrian bridge for two reasons. First to connect the fortress to the city centre, second to make a second escape route from the fortress in case of emergencies.

Concept
In former days, the Ravelijn was supplied from the city with small rowing boats. The concept of the bridge is, to let the bridge follow the original track of these boats, so the bridge echoes the former route the boats followed from the city to the fortress. That is why the bridge snakes across the water to the fortress. Therefore we also made the bridge floating. An additional advantage of that is , that in winter the bridge can be pulled to the side, so there can be ice-skated around the fortress. The deck of the bridge is convex to let the bridge blend in with the water and the surroundings. No mirror image in the water and as close to the water as possible. The stairs at the jetty near the fortress can move up and down (and stay horizontal) with the water level.

Material
The bridge is made completely to the principles of the Cradle to Cradle philosophy. Air-filled polyethylene pipes positioned underneath the timber surface help keep the bridge afloat, without requiring any additional structural framework. The decking is made of Accoya, a high-performance wood product, which is treated to improve its ability to resist fungal decay, and the effect of swelling and shrinkage that could result from its proximity to water. In the future the bridge can be easily disassembled and recycled.

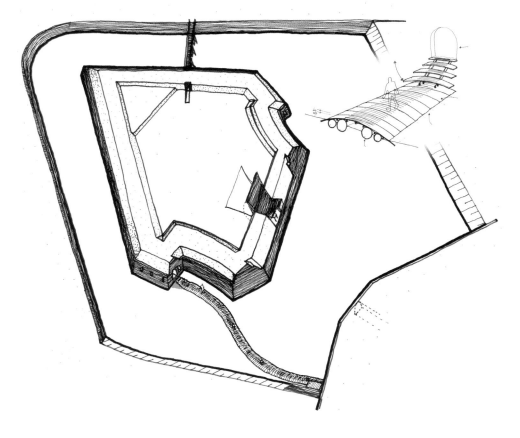

拉维利因堡垒岛"奥普登佐姆"是一座位于荷兰贝亨奥普佐姆市的堡垒-岛屿，它建于十八世纪初，由著名的堡垒建筑师门诺·范·寇恩建造。这也是现存的唯一一个由寇恩建造的拉维利因堡垒岛。最初，这座堡垒只能通过船舶到达，因此士兵必须划船穿越80米宽的水面，将人员和物资送达堡垒。原有入口现在仍旧保留于水岸线正上方。在十九世纪末，堡垒的防御功能被废弃。1930年，水面上架起了一座凸起的木桥。现如今，这座岛屿-堡垒主要用于举行小型的公开或私人的活动。设计任务是为这座堡垒岛屿建造第二座人行桥。原因有二：一是通过新桥连接堡垒与城市中心，二是在紧急情况下，开辟另一条离开堡垒的疏散路线。

设计概念

从前，人们使用小型的划艇从城市为拉维利因堡垒岛运送供给。新设计的这座桥的概念就源自于此，设计师沿着这些划艇原有的行进路线设计了桥的形状，因此桥的路径呼应了从前小船从城市到堡垒的路线。这也是这座桥呈蛇形穿越水面通向堡垒的原因。设计师们让桥浮于

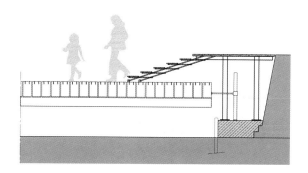
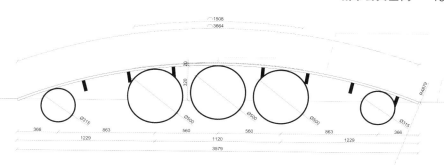

水面。使用这种浮桥获得的额外优势是,在冬天,人们可以把桥拉到旁边,从而在堡垒四周就形成了一座天然的滑冰场。桥的甲板是凸面的,因此桥能融入水中和周围环境中。桥的高度尽可能接近水面设计,因此水中没有桥的倒影。堡垒附近的码头阶梯可以随着水面的高低上升和下降(且保持水平)。

材料

这座桥完全依照"从摇篮到摇篮"的哲学原则来设计。充气的聚乙烯管排布在木材表面的下方,它们有助于保持桥体的漂浮,且不需要任何附加的结构框架。桥的甲板由固雅木制成,这是一种高性能的木制品,作用是提升桥体抵抗真菌侵蚀的能力,并且可以抵消由于桥面接近水面而产生的膨胀和收缩的影响。在将来,桥的拆卸和回收都非常容易。

HAMBURG INTERNATIONAL GARDEN SHOW AQUA SOCCER
汉堡国际园艺展水上足球池

Location: Hamburg, Germany
Completion: 2013
Design: Topotek1
Photography: Hanns Joosten
Area: 1,250sqm

项目地点：德国，汉堡
完成时间：2013年
设计师：Topotek1景观事务所
摄影：汉斯·尤斯登
面积：1,250平方米

The Aqua Soccer Installation at the 2013 Hamburg International Garden Show aspires to take a conventional game and introduce it to new a context. The Aqua Soccer installation offers a fun twist on the traditional sport.

Conventional soccer is played on a flat rectilinear lawn with wide, straight shots made to opposing goal. At the Aqua Soccer installation, the field conditions of soccer are intentionally inverted; everything that was once simple is now a struggle: fast movements are difficult, straight shots are nearly impossible, the goals do not align with each other and the playing field is drastically narrowed. Most importantly, the field surface is rubberised and filled with water.

Playing this revitalised game of soccer becomes a provocative adventure, guided by new rules and free thought. The hybrid game decontextualises soccer into a water sport, bringing humour and transgression to the conventional game. The imaginations of the players are pushed to find new strategies that will allow them to play this new game successfully.

The Aqua Soccer installation pushes the boundaries of traditional play activities in order to revitalise an old game. Lawn-field soccer is a fast game played on flat, rectilinear surfaces, with wide, straight shots made to opposing goals. Water on the lawn can create unintentional comic shots. In Garden 73 these conflicts become the central issues of the programme: fast movements become a struggle, straight shots become difficult. Goals do not stand straight against each other – but are rather scored diagonally. The narrow area of the playing field becomes the centre of the match and must be overcome in order to continue the fight for the ball and eventually score. The playing field is made of a simple concrete basin. The floor is covered with an elastic coating, which is roughened for friction. Flights of stairs serve as both viewpoints of the site, through its slight rising steps, and as a wide sitting area. To the south and west a fence to catch balls bounds the area, preventing them from leaving the playing arena.

2013汉堡国际园艺展的水上足球池从传统运动中获得启发，将其引入了全新的环境。水上足球池是一种对传统运动的大胆颠覆。

传统足球在平坦的草坪上进行比赛，大多以直线球破门得分。在水上足球池，足球运动的场地条件被刻意反转了。曾经简单的运动变得十分艰难：快速移动很难，直线球几乎是不可能的，两边的球门不再是正对的，场地也变得特别狭窄。最重要的是，场地表面采用橡胶处理，并且充满了水。

这种新式足球变成了一种刺激的冒险，需要全新的规则和自由的想法。它将足球变成了一种水上运动，为传统运动带来了幽默感和变化。运动员必须通过充分发挥想象力寻找新策略来取得比赛的胜利。

水上足球池打破了传统运动的极限，赋予了足球这一古老的运动全新的生命。草地足球是一种快速的比赛，在平整的长方形场地上进行，大多以直线球破门得分。草地上的水会导致无心的滑稽射门。在73号园区中，这些冲突成为了项目设计的核心元素：快速移动变成了奋力挣扎，直线球变得异常艰难。两边的球门不再是正对的，而是斜对的。狭窄的比赛场地成为了比赛的核心，球员必须克服这些障碍，不断抢球，才能获得射门机会。比赛场地由简单的混凝土水池构成，地面上铺装着粗糙的橡胶层，以增大摩擦力。场地旁边的楼梯起到了看台的作用，同时也可供人休息。西、南两侧的栅栏能阻挡跳跃的足球，防止足球脱离比赛场地。

KALA – PLAYGROUND AND GREEN SPACE IN BERLIN-FRIEDRICHSHAIN
KaLa弗里德里希斯海因游乐场与绿色空间

Location: Berlin, Germany
Completion: 2011
Design: Rehwaldt Landschaftsarchitekten
Photography: Rehwaldt Landschaftsarchitekten
Area: 3,600sqm

项目地点：德国，柏林
完成时间：2011年
设计师：Rehwaldt景观事务所
摄影：Rehwaldt景观事务所
面积：3,600平方米

Site and Surroundings – The Green Islands
The urban quarters south of Frankfurter Allee are low-density areas with a high green rate ("green islands"). Small squares or green spaces accentuate the district and become sites with high amenity values. The design concept aimed to strengthen these characteristics and to enhance a distinctive identity. The opportunity arose to consider various target groups and different demands respectively for there was a lack in open spaces for children and elderly people. So, a great diversity in sojourn and playing facilities for all age groups could be obtained in the quarter. The idea of two different sites – Meeresinsel (isle in the sea) and Erdeninsel (isle on the earth) came up to deal with the existing deficits. Through a well-balanced arrangement of the elements disturbing interferences could be avoided.

The Meeresinsel (Isle in the Sea) – Play in the Current
The existing penguin playground has been seized and refined in space and issue. The sandy zone has been enlarged southwards into a "big sea". The existing playground equipment were maintained and renewed. The motif of penguins, seals and other sea dwellers was complemented by new playing elements in the extension. The "floes" are varied useable playing objects which pick up the image of sheets of ice which are drifting in the current. Different topics characterise the floes and create various playing situations: climbing on the iceberg or walls, sliding or balancing, "diving" or "marbles". The surfaces are made of coloured concrete which is partly coated. A flying fox is leading from one floe into a sandy area. The playground is surrounded by a low characteristic fence. Large lawn areas were maintained on the intersection to the eastern neighbourhood. Existing shrubs were lightly pruned to improve the vista. But, on the sunny northern edge of the penguin playground the bushes were maintained for screening. A little square on a junction in the south functions as playground entrance. In this paved area the existing water pump as well as a few bicycle racks are arranged.

The Erdeninsel (Isle on the Earth) – A Green Oasis
On the site between Kadiner and Lasdehner street a spacious common area has

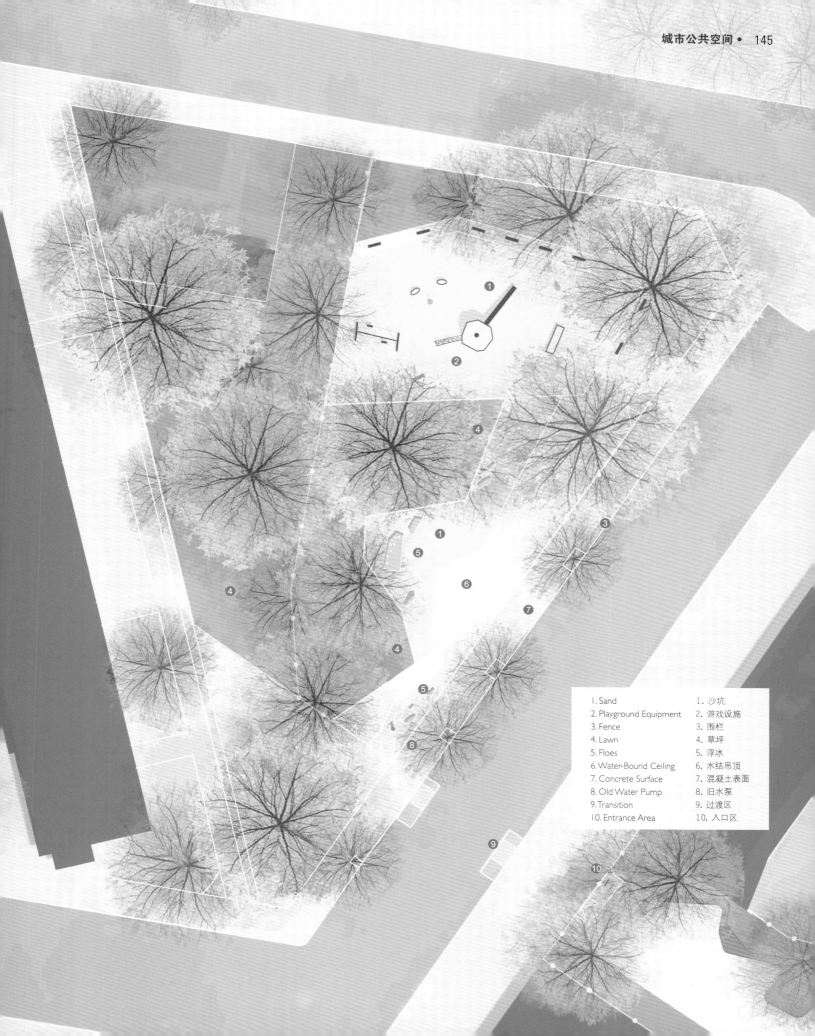

been implemented which allow a self-evident, easy passage. The "Erdeninsel" is a green oasis which complements the active play offers in the neighbourhood. Particularly the adjacent schools are equipped with manifold playing elements. Therefore, a multifunctional space was designed without interfering installations.

The open space is structured by only one feature which gives the space its own identity. The exorbitant "Windenwurm" (winding worm) is bench, reclining area, picnic table and stage at the same time. Long wooden elements wind through the space in elegant curves while they form the border to the adjacent estates at the same time. The vegetation behind was maintained and partly intensified. The undergrowth strengthens the vegetative character and provides the bench a "green background". In front of the southern edge is a small vegetation stripe with lawn and ground-cover plants which attenuates the separative impression of the fence optically. At the same time it connects the space with the adjacent school open spaces.

场地与周边环境——绿色群岛

柏林市法兰克福大道南面的城区是低密度建筑区，绿化程度很高，被称为"绿色群岛"。小型广场和绿地点缀着城区，环境舒适宜人。项目的设计概念旨在强化这些特色，打造独特的形象品质。项目考虑到儿童和老年人缺乏开放空间进行活动，因此为全年龄段的人群打造了丰富多样的休憩和游乐设施。项目通过两个不同的场地来处理预算超支问题，即"海洋岛"和"陆地岛"。合理安排的景观元素将二者巧妙地连接起来。

海洋岛——在水流中游戏

项目对原有的企鹅游乐场进行了空间改造，将沙区向南扩建，形成了"大海"。原有的游乐设施得到了保留和翻新。新建的游乐设施充分体现了企鹅、海豹和其他海洋生物的主题。

"浮冰"是各种各样的游乐设施，其造型就像是漂浮在水流中的冰块。浮冰有不同的主题，营造了变化的游乐环境：在冰山或墙上攀爬，玩滑梯、跷跷板，"潜水"或"鱼跃"。游乐场四周是低矮的特色围栏。大面积的草坪被保留在与东面社区的交叉口上。原有的灌木经过修剪，更加美观。但是，企鹅游乐场北边阳面的灌木丛被留下起到了屏障的作用。南侧交叉口的小广场起到了游乐场入口的作用，在铺装地面上设置着水泵和少量自行车停放架。

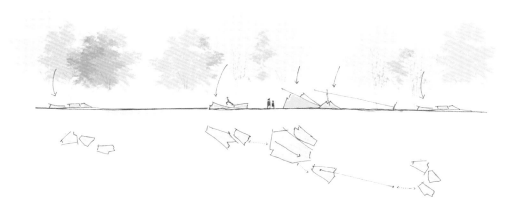

陆地岛——绿洲

在卡迪那街和拉斯德纳街之间的地块，一块开敞的公共区域被改造成了清晰简洁的通道。陆地岛是一片绿洲，由于周边社区的活动十分丰富，特别是附近的学校都配有多样化的游乐设施，因此这片多功能空间并没有设计过多的装置。开放空间简洁而别致，像虫子一样蜿蜒的结构可以是长椅、靠背、野餐桌或舞台。长长的木制构件以优雅的弧线贯穿了整个空间，同时也形成了场地的边界。后方的植物得到了保留和部分改良。树下的灌木突出了植物特色，为长椅带来了"绿色背景"。在南侧边缘的前方，由草坪和地被植物组成的小型植物带在视觉上弱化了围栏的隔离感。同时，它也将空间与附近学校的开放空间连接起来。

SAARBRÜCKER PLACE
萨尔布吕肯广场

Location: Idsteiner Nassau Viertel, Germany
Completion: 2011
Design: Die LandschaftsArchitekten
Photography: Die LandschaftsArchitekten

项目地点：德国，伊德斯坦纳拿索区
完成时间：2011年
设计师：Die景观建筑事务所
摄影：Die景观建筑事务所

In the Idsteiner Nassau Viertel, a town near residential area, the Saarbrucken place is located between new blocks of flats. The generation-covering stay and activity offer and the attractive urbane creation with high-quality designed equipment and individual made elements, like, for example, a pergola, wooden decks and other furniture or fountains create aesthetically and functionally high clearance quality.

The place surfaces almost completely an underground parking garage, but is also practicable for the treacly load traffic. For this static conditions special constructional solutions were required and also big care by the implementation. The adjoining street segment was also developed.

The western place part is marked in its middle by a play of water and seat opportunities like woodendecks and serves the relaxation and the communcation. The place receives its end to the west with three great trees. On a generous open lawn lying and oafs are as possible as plays and raving. To the east the place is framed by a pergola which strengthens the space impression and offers shady seats. The central inside-recumbent surfaces of the place are divided with raised beds for plants a few trees, partially with seat possibilities.

萨尔布吕肯广场位于伊德斯坦纳拿索区新建的住宅区内。老少咸宜的休息和活动环境以及宜人的城市景观设施（如绿廊、木板平台、座椅、喷泉等）共同打造了兼具美观性和功能性的高品质空间。

广场下方几乎全部是地下停车场，但是仍能承受一定的交通荷载。在这种情况下，设计师采用了特殊的构造策略并进行了精心的实施。项目同时还对毗邻的街道路段进行了改造。

广场西部正中是戏水和休息座椅，让人们充分地休闲和交流。广场的西端是三棵大树。人们可以在开阔的草坪上休息和游戏。广场东部有一个绿廊，它突出了空间的特色，也为下方的座椅提供了阴凉。广场的中央空间被架高的树池所分割，树池也可作为座椅使用。

城市公共空间 · 149

城市公共空间 • 151

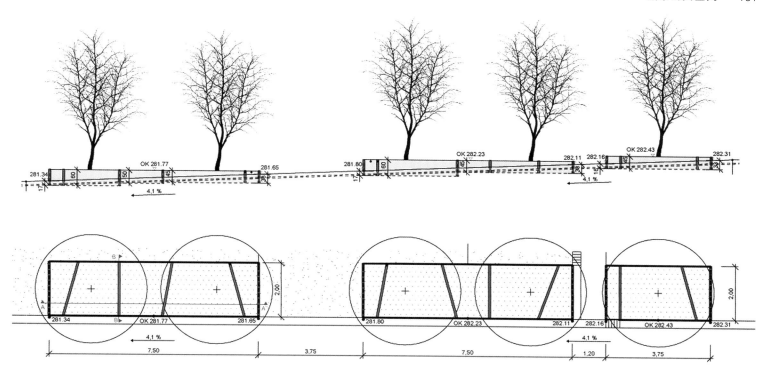

THINK K
K思想广场

Location: Stuttgart, Germany
Completion: 2014
Design: Rainer Schmidt Landschaftsarchitekten GmbH
Photography: Stefan Müller/ Fürst Developments GmbH
Area: 30,000sqm

项目地点：德国，斯图加特
完成时间：2014年
设计师：Rainer Schmidt景观建筑事务所
摄影：斯蒂芬·穆勒/ Fürst开发公司
面积：30,000平方米

Think K, a new mixed-use development area with upmarket architecture, is being developed west of the "Grüne Fuge" green space corridor. While a legible base pattern provides the basic structure for the whole site, it offers the flexibility to allow a varied design character within each of the individual sub-zones of residential, commercial, central plaza and childcare centre. Buildings are grouped around a large green central plaza or "Quartiersplatz". Paths connect the plaza and residential areas directly to the neighbouring "Grüne Fuge" and Killesberg parks. The free-standing nature of the buildings allows an open exchange between the green of the courtyard and the green of the parkland, bringing a park character into the middle of this urban quarter.

Low walls are used as a universal theme of the entire area to present a unified character and to successfully negotiate minor changes of level. The central plaza is identifiable through its three mown grass "fields". These are complemented by a canopy of clipped historic fruit tree varieties. In spring, the fruit tree groves become a giant cloud of blossom and are underplanted with early blooming bulb varieties, and in autumn fruit can be harvested by residents. In every season, the central plaza displays a different character of flowers or foliage. Play equipment, a sand play area, and a graveled outdoor seating zone for a ground-level café are also accommodated in the central plaza. The canopy of trees extends over all of these areas, providing welcome shade from the summer sun.

Via the use of different hedge types and pavement styles, each residential cluster has a recognisable and individual character. Small plazas within the residential clusters serve as central meeting places for the surrounding buildings. Generous garden areas with terraces strengthen the garden-character and ensure a high quality of outdoor space and outdoor lifestyle opportunities.

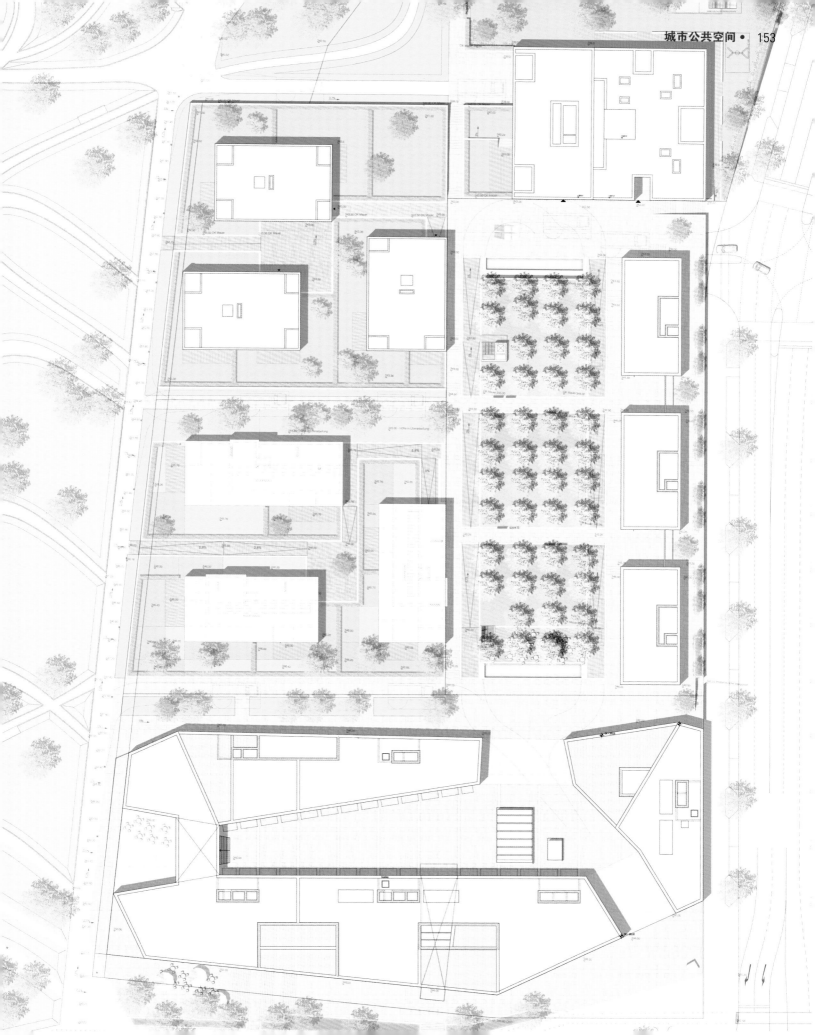

K思想广场是一个拥有高端建筑的混合用途开发项目，建在"青沟"绿色空间走廊的西面。整个项目场地有一个清晰的基本模式，而各个独立分区（建筑、商业、中央广场、儿童保育中心）又各具灵活的特色。建筑环绕着巨大的绿色中央广场展开。道路让广场与居住区直接与附近的"青沟"和基乐斯公园相连。建筑的独立性让庭院的绿地与公共绿地开放交流，为城市街区带来了公园般的感觉。

整个区域以矮墙为主题，呈现出统一的风格，成功地调节了地势的分层。中央广场以三块修剪整齐的草坪为特色，草坪上种植着各种果树。春天，果树林繁华似锦，下方的鳞茎类花草也竞相开放；秋天，居民们可以自行采摘果实。植物的花叶让中央广场的每一季都拥有不同的景色。中央广场还设有独特的沙坑游戏区和碎石铺装地面的露天咖啡厅。树冠延伸到这些区域上方，在夏日带来了无限清凉。

各种树篱和铺装风格的运用让每个居住群都拥有自身的特色。居住群内的小广场起到了中央集会作用。配有平台的花园突出了项目的花园特色，保证了高品质户外空间和户外生活方式。

EL PALMERAL OF SURPRISES
帕尔梅拉惊喜长廊

Location: Málaga, Spain
Completion: 2011
Design: Junquera Arquitectos
Photography: Cemex, Jesús Granada, Heliopol
Area: 6,675sqm

项目地点：西班牙，马拉加
完成时间：2011年
设计师：Junquera建筑事务所
摄影：Cemex、赫苏斯·格拉纳达、Heliopol
面积：6,675平方米

The proposal arises from the search for answers to these three challenges: trying to find explicit references with unsuccessful results, find tips and finally, generate unique models that the architects believe will open up a range of future performances. The architects were looking for a model for a Mediterranean park. After analysing the Arabian garden, all Mediterranean gardens studied had a more domestic character.

Looking, find the reference model farms in the Mediterranean, on the Draa River Valley in Morocco and in the few remaining in Elche, in which the palm is the crop structure element in 20m x 20m grids, and the palm plantations having separations of 3m. The size of these grids is the result of the millennial experience, creating a microclimate protected against sun and wind necessary for subsistence crops in a climate as hard as North Africa. This grid provides a sequence of spaces bounded by the stems of palms and their tops, forming empty scenic spaces treated as different gardens, agricultural plantations replacing uses for the stay, leisure or recreation, clothed with plantations of different size, texture, flowering period, smells and colour. This model makes the Park spatial order as both a sub sequence diversity of different activities spaces and under a continuous blanket shade reinforced by a second order generated by the various plantations which are arranged inside the grids.

The Double Condition of Garden and Promenade
After solving the model garden is necessary over the shadow that must accompany the walk. Sunny in winter, shade in summer. The architects designed a structure that met this condition generating the walker clothing shadow along the 500m of the ride. An undulating canopy sequence similarity of awnings protects the streets of Toledo, Seville and Malaga's Calle Larios. A high level of irregular trace generates a space that is perceived as a large vault. The architects designed a mixed structure of steel beams and pillars decomposed into slender small pillars like bamboos, to convey an image of lightness and transparency despite its important structural conditions, 11m high, 25m apart. The steel beam hangs slats sequence traces varied 14m in length, 18cm in thickness and drape varying between 20cm and 2.50m.

158 • URBAN PUBLIC SPACE

项目方案对来自三方面的挑战进行了探究：找到不成功案例的明确参考，找到窍门，生成独特的模型来实现未来的绩效。建筑师试图打造一个地中海式公园。在对阿拉伯园林进行分析研究之后，他们认为地中海园林应当更具西班牙本土特色。

设计师将摩洛哥德拉河山谷中以及西班牙埃尔切市的地中海农场作为参考模型。作为田地的分隔元素，农场的棕榈树以20米×20米的网格形式呈现，每棵棕榈树的间隔为3米。网格的规格经历了上千年的实践，为农作物提供了能够抵御北非日晒和风雨等恶劣天气的微气候环境。在本项目中，这种网格通过棕榈树的树干和树冠提供了空间序列，形成了空白的景观空间。农作物被休闲娱乐的花园所取代，点缀着拥有各种规格、质感、花期、香气、色彩的植栽。这个模型让公园的空间序列兼具多样性和统一感，既拥有多样化的活动空间，又有一个统一的空间网络。

花园与散步长廊的双重身份

在解决了模型问题之后，冬暖夏凉的散步环境就成为了项目的主要议题。设计师设计了一个能为步行者提供遮阳避风的500米长的散步长廊。波浪形的华盖与屋檐相似，将托莱多街、塞维利亚街和马拉加的拉里奥斯街保护起来。不规则的轨迹所形成的空间像一个巨大的拱顶。设计师设计了一个钢梁和立柱混合结构，就像是多根竹子捆绑在一起，显得轻盈而通透，立柱的高度为11米，间距25米。钢梁板条的长度为14米，厚度18厘米，向下垂下的距离在20厘米至2.5米之间。

GARDEN OF THE SILHOUETTES
剪影花园

Location: Altea, Spain
Completion: 2013
Design: Esculpir el Aire
Photography: David Frutos

项目地点：西班牙，阿尔特亚
完成时间：2013年
设计师：Esculpir el Aire景观事务所
摄影：大卫·弗鲁托斯

The Main Door
A certain point of time exists in the architectural work definition where the outdoor space skin and the indoor space skin come together: the main door. There, the haptic aspects of sight are at its height. Rhythm slows down and the feet then seem to pause, the man's skin and the building's skin touch each other. It's the most intimate moment of the encounter between Man and Architecture.

Rhythm Is Also a Tactile Experience
The skin breakup produces rhythm. All around the perimeter of the Garden's skin it is taking place the multiplicity of "the silhouettes". For this, it has been used a series of small ceramic pieces in vertical format (10x20cm) with different colours in a cold range. The top of "the blue ribbon" is made of squared ceramic pieces, in a 20x20cm format, offering a more static perception. The bottom of the façade is made up of vertical, white ceramic pieces in 10x20cm format: it's the same texture that "the blue ribbon" but with no colour, so that it reinforces the perception of the inclination that the street shows. When the night comes down, the Garden should not lose its rhythmic character. The artificial lighting of that space offers the same vibrant, sequenced character throughout the night just like the day, by using thin luminaries inside the vertical joints of the ceramic pieces.

The Material Skeleton of the Garden of the Silhouettes
A space for children under three years old means, unavoidably, to think of the scale of the built space in future. So, the Garden of the Silhouettes operates in front of three gradual levels: the scale of a child, the scale of Man and the scale of the building and the environment. This process of manipulation is especially necessary because "an architecture of infinite space is not understandable, because it exceeds the system of perception, the structures of perception." If "contour and profile are the cornerstone of the architect", the floor surface should take back the main role on the stage that it deserves: It's there where "the dancer" moves, "who has his ear in his toes". The important thing is not to identify the surfaces of that one surrounds us, "but to perceive the character of that one surround us".

正门

正门代表着内外墙面的临界点。实现的触感在这里达到了最高点。节奏慢了下来,脚步也似乎停止,人与建筑的皮肤相互触摸。这是人与建筑的接触中最亲密的一刻。

节奏同样是一种触感体验

外墙的分裂产生了节奏。花园外墙的表面呈现为多样化的剪影。各种各样的小尺寸瓷砖(10x20厘米)垂直排列,呈现为各种冷色调的条纹的组合。"蓝缎带"的上方墙面贴有20x20厘米的方形瓷砖,给人以更宁静的感觉。墙面下部由10x20厘米的白色瓷砖构成,纹理与上方的"蓝缎带"相同,但是没有颜色,突出了街道的斜坡感。即使夜幕降临,花园也不会丧失自身的节奏感。潜在瓷砖垂直接缝内的细灯管为其带来了与白天一样的生机和序列感。

剪影花园的材料骨架

专为三岁以下儿童所设计的空间无可避免地要考虑到未来建造空间的比例。因此,剪影花园的运营面临三个渐变的层次:儿童的比例、成人的比例、建筑和环境的比例。空间处理的过程十分必要,因为"无限空间的建筑是不可了解的,因为它超越了感知系统和感知结构"。如果说"外形和轮廓是建筑师的基石",地面则是舞台的主角:作为"舞者"舞蹈的地方,它的触感至关重要。设计的重点并不是界定出环绕我们的界面,而是"感知它们的特点"。

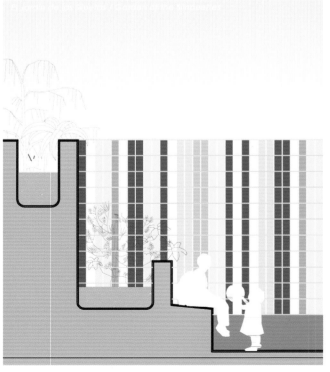

RICARD VIÑES SQUARE
理查德·韦恩斯广场

Location: Lleida, Spain
Completion: 2011
Design: Benedetta Tagliabue
Area: 9,200sqm

项目地点：西班牙，莱里达
完成时间：2011年
设计师：贝娜蒂塔·塔格利亚布
面积：9,200平方米

The large green open spaces that surround the Seu Vella Cathedral and dominate the whole city are the most beautiful public areas in Lleida, and are what the architects chose as the reference point. The designs for the new Ricard Viñes Square must possess some of this beauty. The focus of the proposal is to build a large green open space for a sculpture dedicated to the musician Ricard Viñes. A space full of little squares and green areas at a point where the city throngs with traffic and pedestrians. The maze or labyrinth provides an ancient model. The cultural meaning and interpretation of the symbol of the labyrinth run deep. The origin of the "labyr" part of the word has to do with rocks and stone, while the "inth" comes from a Greek word meaning foundation site. Even though the precise etymology is unclear, there was a time when path mapping was used to notate dance choreography.

The architects propose an open space featuring a dance floor with a labyrinthine path guiding the steps of those dancing the spring dance around the central feature – a feature that generates and guides the movement of the dance, filling the surround space with life. People and traffic will move differently in Ricard Viñes Square, where pedestrians will own the public space.

Everyone will follow the paths mapped by the dance notation, but with no overlap. Perhaps the closest reference to the design idea is Francesc Macia Square in Barcelona, in one of the busiest areas of the city, and at a key transit point in and out of the city. Other European cities are famous for the spectacular circles in their metropolises, such as the Etoile in Paris or Berlin's Tiergarten with its famous column topped by an angel. In the UK, the architect John Wood created one of Bath's most iconic features with the Royal Crescent whose circular green spaces control the flow, and substantially contribute to defining and propagating the city's image. Circular spaces are not normally accessible and for this reason are often laid out as parks, regulating traffic flow and giving drivers a glimpse of the natural world.

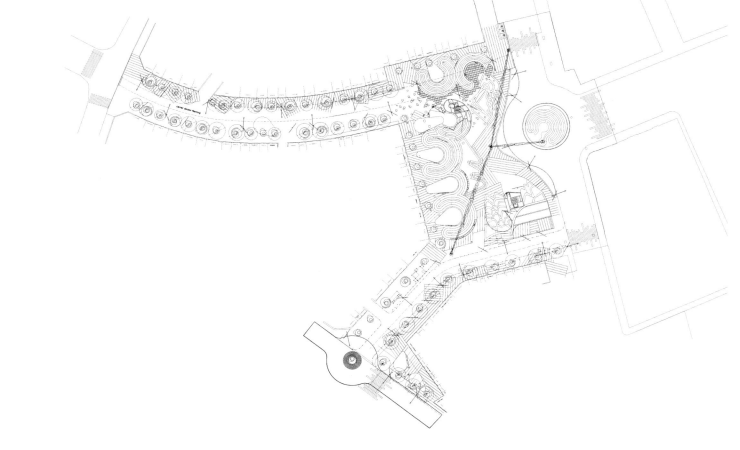

环绕塞乌维拉大教堂的大片绿地是莱里达最美丽的城市公共区域，也是项目设计的参照物。新理查德·韦恩斯广场的设计必须具备这种美好的品质。设计方案的重点是以著名音乐家理查德·韦恩斯的雕像为中心，打造一片开阔的绿色空间。整个空间由小型广场的绿地构成，被熙熙攘攘的城市人流和车流所环绕。迷宫式的设计显得十分古老而神秘。迷宫（labyrinth）象征的文化内涵十分深邃。英文单词迷宫的前半部分"labyr"与岩石有关，而后半部分"inth"则来自于希腊语中的"创建地"。尽管它的确切词源并不确定，人们曾经利用路线图来表示舞蹈编排。

建筑师设计了一个以舞池为特色的开放空间，迷宫小路引导着跳着春天舞曲的人们环绕中央景观元素舞蹈。中央景观元素引导着舞蹈的运动，让空间充满了生机。人流和车流在理查德·韦恩斯广场上以不同的方式移动，行人拥有自己的公共空间。

人们随着舞蹈编排的路线移动，但是路线不会重叠。项目明显参考了巴塞罗那的弗兰塞斯克·马西亚广场的设计，后者同样位于繁忙的城区中，在进出城的主要经停点上。其他欧洲城市大都以城市中壮观的圆圈地带而著称，比如巴黎的星形广场、柏林的蒂尔加滕公园等。在英国，建筑师约翰·伍德打造了巴思最著名的皇家新月楼，它的圆形绿地控制着人流，对塑造和宣传城市形象做出了重大的贡献。圆形空间不仅便于进出，而且还能以公园的形式展开，控制交通流量，让驾驶者可以一瞥自然世界的美景。

HYLLIE PLAZA
希里广场

Location: Malmö, Sweden
Completion: 2011
Design: Thorbjörn Andersson
Area: 14,000sqm

项目地点：瑞典，马尔默
完成时间：2011年
设计师：Thorbj rn Andersson景观事务所
面积：14,000平方米

No tree is more characteristic of Scania, Sweden's southernmost province, than the beech – and no other species bears such a strong association with its home region. The horizontal branch structure that supports its canopy, paired with the equally horizontal orientation of its leaves, produces a quality that is almost architectural. The trunk, with its smooth, light-grey bark, in some ways resembles the leg of an elephant. As light filters through the beech canopy its leaves appear almost transparent, with delicate colours that add to the feeling of freshness. Further into the year, its leaves become hard and leathery, which in turn renders a sharp contrast between the tree's areas of light and shadow.

The competition entry for the new plaza in the Hyllie district of Malmö bore the motto Fagus, the scientific name for beech. The idea driving the design was to establish a beech forest on the plaza, or the plaza as a beech forest. Being a characteristic species for the region the forest would contribute a regional identity to a site almost without character. The beech tree, then, would be the trademark for the new plaza. Later the designers were to learn, however, that the design concept was as ignorant as it was unique: the beech doesn't thrive in the conditions provided by the site – it would require help. In nature, the growth of the tree depends on free soil with an active exchange of oxygen and carbon oxide at its roots. It desires mulch in a perpetual turnover and benefits from the falling organic debris characteristic of the forest. It is sensitive to drought. Despite these challenges the motto for the competition entry was set and the aim of the proposal at this stage was no longer negotiable. In response significant biological research began, a project that included the efforts of many experts trying to find a high-tech solution for ensuring the life of the Hyllie beech grove.

First, came a gigantic planting bed that equalled the dimensions of the plaza above it. This earthen layer consisted of an 80 cm- thick base course of football-sized boulders that form a structural soil, 60 % of which is cavities. This volume was filled with mulch which was watered down once the boulders were in place. The planting bed was capped with Swedish high-density granite paving totalling

12,000 square metres in area. Each paving stone measures 2 x 1 metres and is 12 cm in thickness, with sawn sides and a thermal finish. They are laid out with exposed joints in both directions. Into this granite field, twelve parallel slits were cut, each planted with two, three, or four beech trees. The earth in the planting bed in each slit has been mixed with pumice and mycorrhiza: Pumice is petrified lava ash with an outstanding capacity to retain moisture; mycorrhiza is a living mushroom/root mixture that improves the tree's nutrient turnover through symbiosis.

By their placement the trees form a series of glades on the plaza. Within the glades seating of various kinds rests on surfaces paved with blocks of wood. Eleven masts, each sixteen-metres tall, are grouped as pairs along the sides of the plaza and frame its boundaries. Between the masts stretch 1,800 metres of steel cables, arranged in a certain disorder resembling a spider's web. The cables support a field of 2,800 LED diodes programmed to four seasonal scenarios that create a digital sky after dark.

城市公共空间 • 173

最能代表瑞典最南端省份斯肯尼亚省的树木要属山毛榉，没有任何物种比它与这一地区的联系更加紧密。支撑树冠的横向树枝以及同样横向生长的树叶营造出一种接近于建筑的品质。光滑的浅灰色树干在某种程度上就像象腿。当光线透过树冠，树叶几乎变得透明，它们淡雅的色彩给人以清新之感。随着时间的增长，树叶会变得坚硬而有韧性，与树木柔和的光影效果形成了强烈的对比。

马尔默市希里区新建广场的获奖设计方案名称就是"山毛榉"。设计方案的出发点是在广场上打造一片榉树森林，或者说是让广场成为榉树森林。作为当地最具代表性的物种，榉树森林将赋予广场地域特征，而山毛榉则会成为新广场的商标。但是后来设计师意识到这一设计概念是不可行的：山毛榉根本无法在场地环境中茁壮成长，它需要借助外力帮助。事实上，树木的生长取决于根部的疏松土壤，土壤中的氧气和二氧化碳能够实现活跃交换。它需要不断翻转的护根物，能够从森林的落叶有机碎片中获益，并且对干旱十分敏感。尽管面临如此多的挑战，项目的基本设计方案已经确定下来，无法改动。因此，项目进行了大量生物研究，许多专家通力合作，共同寻找能让山毛榉在广场上生长的方法。

首先，在广场地面上设计了一个巨型种植池。种植池内的土层由80厘米厚的岩石（足球大小）基层作为结构土壤，60%的空间是中空的，用护根物填充。种植池上方的广场地面铺有高密度瑞典花岗岩，总铺装面积可达12,000平方米。每块石材的尺寸是2×1米，厚度为12厘米，配有锯切侧面和保温饰面。石板的四面都有接缝。在花岗岩地面上，切割出12个平行切口，每个切口内种植2~4棵山毛榉。种植池切口内的土壤混合了浮石和菌根：浮石是石化熔岩灰，具有卓越的保湿性；菌根是活菌，能够通过共生定期为树木补充营养。

这一设计让树木在广场上形成了一系列的林中空地。设计师在空地上设置了各种座椅。11根16米高的桅杆被放置在广场两侧，标志出边界。桅杆之间拉伸着1,800米的钢缆，像蜘蛛网一样错乱。钢缆上点缀着2,800个LED二极管，通过编程，它们会在夜晚营造出闪亮的天空。

TALENTENCAMPUS VENLO
芬洛人才校园

Location: Venlo, the Netherlands
Completion: 2012
Design: Carve
Photography: Carve
Area: 7,000sqm

项目地点：荷兰，芬洛
完成时间：2012年
设计师：雕刻景观设计事务所
摄影：雕刻景观设计事务所
面积：7,000平方米

Carve designed the outdoorspace of the 'Talentcampus', a new type of school that combines closed, half-open and normal primary education. Instead of focusing on their limitations, the talents of the children are stimulated. The three schools are integrated in one building – there is no strict separation between them, which enables a fluid communication and exchange. This principle was also the starting point for designing the surrounding schoolyards. The challenge was to offer the diffent groups a space of their own, whilst encouraging interaction between them at the same time. We found the solution in creating squares like compartments. By opening or closing the fences between them, the size of the playgrounds can be either enlarged or reduced, depending on the size and type of the group of children. The 'stages' around the school are a buffer between the play- and parkingzone. They are not only framing the play, but are also a sitting element and storage space. On the northside, a slanting concrete wedge, on one side clad with corten-steel, sticks out of the playground like a tectonic plate. The elevation is sitting edge, frame of the basketballcourt and backstop for the bike parking – but also a buffer to the street. The most challenging aspect of the project was to formulate an answer to the demands on parking and water infiltration on site. The large amount of parking spaces had to be solved on the edges of the schoolyard, but this would result in a very large paved surface. The wish to soften the whole led to the invention of a new 'grass-brick', the Greenbrick, which was developed especially for this project.

Greenbrick

'Green' and infrastructure are a difficult match; they are two different entities. The urban environment is getting more 'stony' though and, as a result of this, the problems with infiltration of rainwater more acute. Attempts to implement grass stones often fail. Standard grass stones are not easy to combine with attractive looking brick pavement, and they often don't meet the technical requirements of a road used by traffic. The invention of the Greenbrick is our answer to this problem, with which we were confronted in the project Talentcampus Venlo. The urban plan asked for solving the heavy parking demands within the boundaries of the site, whilst at the same time it was an explicit wish to make the area more green and attractive.

176 • COMMERCIAL & INSTITUTIONAL SPACE

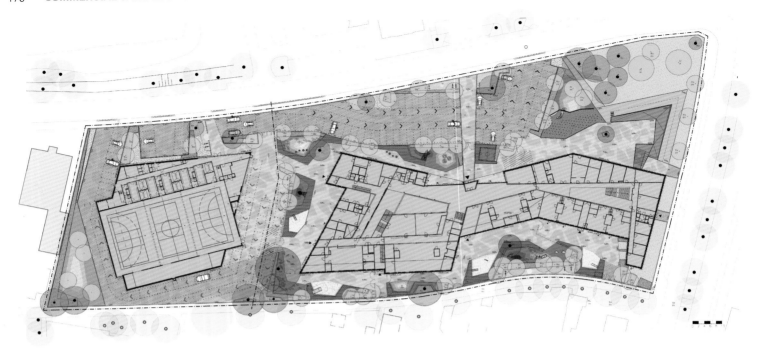

雕刻景观设计事务所承接了芬洛人才校园的户外空间设计项目，人才校园是一所综合了封闭式、半封闭式和普通初等教育三种教学模式的学校。设计师没有被这种局限性束缚，而是将设计重点放在激发孩子们的才能上。三所学校的教学活动集中在一座建筑物中，学校之间没有严格的界限，这种设计方式使流畅的沟通和交流成为了可能。流畅沟通与交流的原则，也是设计师在设计校园环境时的出发点。设计师面临的挑战是，不同群体需要各自的空间，而这样的空间同时又应有助于不同群体之间的互动。设计师的解决方案是创造了像隔间一样的方形空间。依据使用学生群体的规模和类型，通过打开或关闭广场间的围栏，操场的尺寸可以被放大或缩小。在学校四周的游乐区和停车区之间的"舞台"起到了缓冲的作用。它们不只框定了游乐区，同时也可作为座椅小品和存储空间。在北侧有一个斜向的混凝土楔子，它的一面覆盖着柯尔顿钢铁，这个楔子如同地壳板块一样，突出于运动场地面。篮球场地的边缘是抬高的，可供就座，也可为自行车停车提供支撑——同时这些边缘也是校园与街道间的缓冲区。这一项目最具挑战性的部分是，设计师需要在满足停车需求的同时，保证场地良好的雨水渗透。大面积地停车场地必须布置在校园的边缘，但是这将会带来大面积的铺装表面。为了实现柔化地面的目标，设计师特别为这一项目开发了一种新型的"草坪砖"——这项发明叫做"绿砖"。

绿砖

"绿化"与基础设施很难两全；它们是两种不同的实体。城市环境变得越来越"坚硬如石"，这种现象带来的结果是，雨水渗透的问题越发严重。人们试图使用草坪砖增加绿化的方法通常不能获得期望的效果。普通的草坪砖不易与人行道美观的砖石铺装结合，它们通常不能满足道路的通行技术要求。设计师在芬洛人才校园项目中发明的绿砖解决了这个难题。城市规划要求使用基地的边界解决大量的停车需求，同时明确希望这一区域能有更多的绿化空间，更具吸引力。

INTERNATIONAL SCHOOL EINDHOVEN
艾恩德霍文国际学校

Location: Eindhoven, the Netherlands
Completion: 2013
Design: Buro Lubbers landscape architecture and urbanism
Architect: Diederendirrix architecten
Photography: Buro Lubbers landscape architecture and urbanism
Area: 95,000sqm

项目地点：荷兰，艾恩德霍文
完成时间：2013年
设计师：布罗·吕贝尔斯景观建筑和城市规划设计事务所
摄影：布罗·吕贝尔斯景观建筑和城市规划设计事务所
面积：95,000平方米

From militairy compound to international school campus

The Constant Rebeque Barracks, a military compound, originally focused on order and discipline, ranking and rating. Fairly far from the city the hierarchical world of men was rather secluded and isolated of society. Since the barracks were no longer used as such, the complex and its surroundings are transformed into the International School Eindhoven (ISE). The typical military atmosphere is changed into a modern and vibrant school campus where hospitality, friendship, meeting, improvisation, participation and the ISE-educational concept blossom. The multidisciplinary team realised an integrated design. The key concept of the design is: connection between scenery spaces, between architecture and landscape, between people. Buro Lubbers was responsible for the design of landscape and outdoor space.

The essence of the landscape design is the division into separate 'rooms': an urban cluster around the school buildings, a space for sports and a more natural space of trees and heath. These spaces are separated by existing trees, supplemented by autochthonous species such as beech, birch and oak. Furthermore, the design is based on symmetry axes which already characterised the military compound. The original ax that connects the gatehouse with the parade ground surrounded by buildings is kept at the new campus and emphasized by creating a new perpendicular axis. This new axis connects the school buildings with the sport fields and the wood cluster. Moreover, this axis forms the new gateway to the campus.

The result is a clear framework in which programmatic changes can take place while the spatial quality of the campus primarily improves. Functions can be added in longer term, temporary initiatives are also possible in short term. The framework offers space for sporting, playing, camping, ice skating and events.

商业与综合景观 • 179

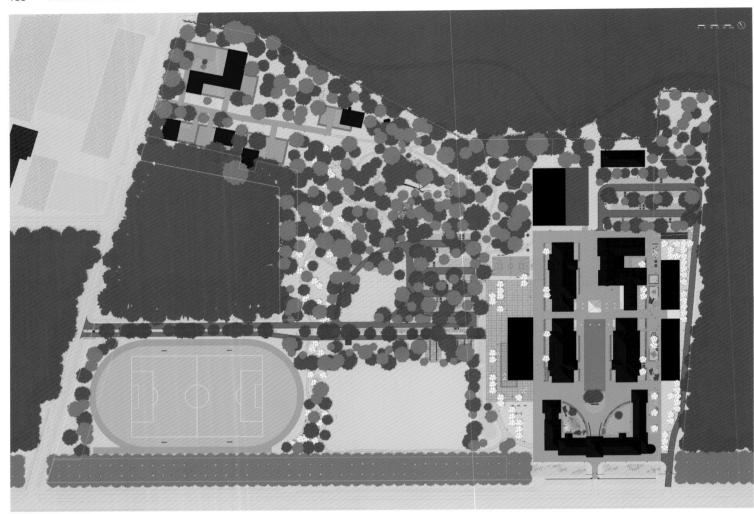

从军事基地到国际学校的转化

永恒瑞贝克兵营原本是一个军事综合体建筑，它强调规则和纪律、等级与身份。这个与世隔绝的隐蔽地方的氛围与城市截然不同，这里的人们等级分明。但是自从兵营废弃之后，这个综合体建筑及周围的区域便被改建成了艾恩德霍文国际学校（ISE）。原本典型的军事氛围，转变成了现代化、充满活力的校园气氛，人们迅速建立起了好客、友善、喜欢集会、热爱即兴创作、重视参与和分享的氛围，ISE的教育理念得以实现。多学科的设计团队对校园的整体方案进行了设计。设计的基本概念是：打造景观空间之间、建筑与景观之间、人与人之间的联系。布罗·吕贝尔斯负责景观和室外空间的设计。

这个项目景观设计的本质是，将基地分割成不同的"空间"：在兵营附近建立起的建筑群，作为学校的各种功能性建筑，另外还建设了一个运动场地和一个气候湿润且有益健康的自然空间。这些空间被原有的树木分隔开来，设计师还补植了一些当地的植物，例如山毛榉、桦树、橡树。此外，设计以原有对称轴为基础，这些对称轴刻画出兵营的刚毅特征。原本连接着警卫室和练兵场的轴线被建筑群包围，在新校园的设计中，设计师将它保留了下来，并增加了一条与之垂直的轴线，来强调这条原有的轴线。新增的轴线，将学校的建筑物与运动场、森林连接起来。此外，这条新的轴线也成为了通往校园的新的路径。

设计的结果是，新的结构改变了场地的原有规划，校园的空间特征从根本上得到了改善。从长远来看，可以为其增加更多的功能，而短期内可以采取暂时性的措施。校园的结构为运动、游戏、露营、滑冰和举办活动都提供了空间。

SCIENCE PARK AMSTERDAM
阿姆斯特丹科学公园

Location: Amsterdam, the Netherlands
Completion: 2014
Design: Karres en Brands landscape architecture + urban planning
Photography: Karres en Brands
Area: 540,000sqm

项目地点：荷兰，阿姆斯特丹
完成时间：2014年
设计师：卡勒斯和布兰茨景观建筑+城市规划事务所
摄影：卡勒斯和布兰茨景观建筑+城市规划事务所
面积：540,000平方米

Science Park Amsterdam is a new type of university site. It is situated in a rather isolated spot on the edge of the city, bounded by roads and railways, water and greenery. This location in the urban periphery therefore does not fit with the idyllic image of the campus as an academic enclave in the landscape, but nor does it match the image of the university as a hallowed institution in the heart of the City. It is to a large extent a self-supporting entity for which a new spatial concept has been devised.

The historical polder layout provides a straightforward template for this expansion, with flexibility for variation and future developments. Clear zoning is introduced over the entire site, creating a distinction between the built – up areas and the open structure in the polder strips. In this way the area is subdivided into five bands of development separated by four green, visual corridors and bounded by green edges. This spatial scheme is not only well suited to the location but also makes feasible a simple logistical organisation. The corridors are used for delivery and dispatch, and provide space for pipes and water features.

The spatial and social cohesion of Science Park consists of a system of semi-public meeting places in and between the buildings. This layout is dissected by a clearly recognisable network of paths, which is only open to low-speed traffic and delivery vehicles. The network of public spaces acts like a nervous system that can expand and contract, depending on the time of day. During the day the campus is lively and busy, in the evening it is quieter, and at night the grounds will for the most part be deserted. The spatial design is in tune with this rhythm. During the day all the neuron bundles are functioning ; at night just the essential systems.

Setting out from the main routes, the visitor proceeds via the public and semi-public spaces in and around the buildings to his or her eventual destination. During the day, all these spaces are open and accessible, establishing a many-branched network. In the evening, the quiet sections are closed off so that the arterial routes, serving functions such as sports facilities, cultural amenities, a hotel and a

conference centre, remain lively and safe. It forms an autonomous network of functions and amenities that, in turn, is tacked onto the greater urban network of Amsterdam and the Randstad conurbation.

The spatial layout also establishes a clear-cut framework for the architectural infill. There are no prescribed alignments and architectural volumes, but a number of development rules do apply. These make it possible to react to changing circumstances and provide some latitude for different architectural ideas. The buildings are grouped in ensembles of different volumes, each ensemble with a single accent in height. The materials for the buildings and public spaces must be pure and should be used in their natural colours. This ensures cohesion in the urban design while retaining architectural freedom.

阿姆斯特丹科学公园是一种新型的大学校园景观。公园坐落于城市边缘，位置相当偏僻，它的四周被公路、铁路、水体和植被包围。由于这里属于城市的外围，因此公园既没有给人以风景环绕的田园印象，也不与城市中心的神圣的校园形象相符。在很大程度上，它是一个自给自足的存在，它开创了一种新型的空间概念。

历史悠久的围垦地布局为公园此次的扩建提供了简单明确的模版，也为未来多样且灵活的开发提供了可能。设计师对整个基地进行了明确的分区，在带状围垦地上的建筑区域和开放结构之间创造出明显的区别。通过这种方式，基地被分成五个带状开发区域，它们之间被四条绿色的视觉通廊分开，基地的四周由绿色的植物划定出边界。这样的空间结构组合方式不只与基地条件相得益彰，同时也形成了一个可行的、简单且有逻辑的组织方案。廊道用于运载和调度，同时也为管道和水景提供了空间。

位于建筑内部和建筑之间的半开放式集会空间系统将科学公园的空间属性与社交属性结合了起来。平面的布局被清晰可辨的道路网络分割开来，道路只对低速交通和运载车辆开放。根

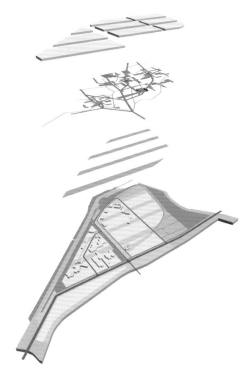

据一天中的时间不同，公共空间的网络像神经系统一样伸展开或收缩。在白天，校园热闹而忙碌，而在傍晚，这里更安静，到了夜里，大部分的场地空无一人。设计师在进行空间设计时，保持着与这种节奏的一致性。在白天，所有的神经元组都在运作；到了晚上只有必要的系统在运行。

由主干道出发，游客们穿过建筑内部和建筑周围的公共、半公共空间，到达最终的目的地。

白天所有这些空间都是开放且可以进入的，它们建立起了一个有着许多分支的网络。在晚上，场地安静的部分将被关闭，因此仅有通达服务功能区——例如运动设施、文化设施、宾馆和会议中心——的主要路线仍旧保持着活力和安全。它形成了一个自发的功能便捷的网络，进而连接至更大的、阿姆斯特丹和兰斯台德卫星城的城市交通网络。

这种形式的空间布局也为建筑的填充建立了一个轮廓清晰的框架。设计师没有对平面图和建筑体积做出规定，但是它适用于许多建设规则。这样的布局使得对环境的改变作出响应提供了可能，为不同的建筑理念的自由发挥提供了空间。设计师根据建筑整体体积的不同对各栋建筑进行分组，每个整体都有各自不同的高度特点。设计师在选择建筑和公共空间的材料时认为，材料必须纯粹且应该使用它们的天然色。这样的设计理念，在保留了建筑风格自由的同时，也确保了城市设计的聚合力。

POLICE FIRE BRIGADE APELDOORN
阿培尔顿公安消防局

Location: Apeldoorn, the Netherlands
Completion: 2011
Design: LODEWIJK BALJON landscape architects
Photography: LODEWIJK BALJON landscape architects

项目地点：荷兰，阿培尔顿
完成时间：2011年
设计师：罗德维克·巴廉景观建筑设计事务所
摄影：罗德维克·巴廉景观建筑设计事务所

The new accommodation of police and security region North and East Gelderland is embedded in a leafy and shady setting that is characteristic for Apeldoorn. The green character of the main road is emphasized with a spacious grass embankment, that withdraws the parked cars from view. The deep orange-brown coloured Corten steel provides a nice contrast with the green grass.

The beautiful tree population of monumental beech and the historical oak woods have been the basis for the design. The two buildings of the police station and fire brigade sprawl among the trees. The configuration of the buildings together with the landscape design not only ensure that all the trees can be kept, but also provide an exciting interaction. Thus creating a natural balance between structure and nature.

The striped pattern of bamboo hedges and grass are repetitious elements creating a calm and continuous carpet for the majestic trees. The light green leaves and the yellow stem of the bamboo give contrast in the shade. Long lines of polished black nature stone reflect the sky and cast daylight under the canopy of the trees. In the evening, lighting, incorporated into the nature stone lines and benches, add to the pattern of lines. The design for the car park is sober; a broad field with lighting poles enlivened with a few trees. The eastside of the area is defined by an artificial stream. This stream is now again located at the original, historical location. The environmentally friendly designed area around the stream and the existing monumental trees keep the area integrated into the ecology of the surroundings.

这座新建成的住所，位于海尔德兰东部和北部的治安与安全区，这里是阿培尔顿典型的多叶和多沙的环境。为强调出主干道的绿色特征，设计师设计了一个广阔的草坪路堤，路堤也将停泊的汽车遮挡起来，隐藏了不雅的景象。深黄褐色的柯尔顿钢则与绿色的草坪形成了和谐的对比。

设计以美丽且古老的山毛榉和橡树群落为基础。警察局和消防队的两座建筑在这些树木中延伸。设计师将建筑形态和景观的设计很好地结合起来，通过这样的设计，不只保证了所有的树

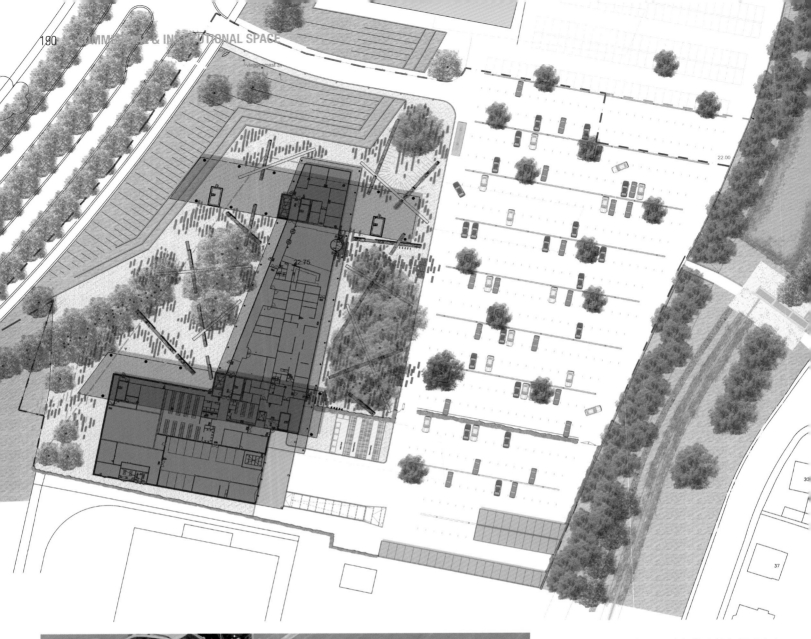

木都可以保留下来，同时也使建筑与景观之间产生了令人兴奋的相互作用。从而创造了建筑结构与自然环境之间的生态平衡。

竹篱笆和草坪的条纹样式不断重复，如同为庄严肃穆的树林铺上一条平静且连续的地毯。浅绿色的树叶和黄色的竹子茎干，形成了颜色深浅不同的对比。地面上有许多长条的黑色抛光自然石，它们反射出天空的图案，又将太阳光投射到树木的华盖之下。在晚上，灯光加上条形自然石和长椅，增强了这种线条感。停车场的设计朴素；广阔的场地上，几棵树在灯柱的照射下熠熠生辉。这一区域的东侧边界是一条人造溪流。溪流现在的位置正是在它历史上曾经存在过的地点。设计师在对溪流周边区域的设计时，注意保护其生态环境，并将古老的树木保存下来，通过这些措施，这一区域与周围的生态环境融为一体。

BEUKENHORST ZUID HAARLEMMER-MEER
哈勒默梅尔勃肯霍斯特南区

Location: Hoofddorp, the Netherlands
Completion: 2014
Design: LODEWIJK BALJON landscape architects
Photography: LODEWIJK BALJON landscape architects

项目地点：荷兰，霍夫多尔普
完成时间：2014年
设计师：罗德维克·巴廉景观建筑设计事务所
摄影：罗德维克·巴廉景观建筑设计事务所

Beukenhorst-South is the latest office park in the Haarlemmermeerpolder. Next to the railway station and close to Schiphol airport the area has good potential to contribute to a sustainable development. The main artery consist of an avenue: functionally because of its central position, and spatially by its central verge that runs as a grassy dike with a double row of Fraxinus trees. Along the avenue runs a wide canal, another strong structural element of the polder landscape. Because these long lines continue in the surrounding, the development is well anchored in the landscape.

The particular elaboration of the steel sheet piling along the canal, indicates that the polder has transformed into an urban area. The steel sheet piling is partly covered with planting, creating an attractive alternation of rust brown stripes and vegetation. It is also an expression of the sustainability ambition.

Over the various polder canals different bridges form characteristic elements in the landscape, and offer a fine fabric of pedestrian routes. The characteristic Corten Steel Bridges cross the canal. In materialisation they connect to the toughness of the quay, in detailing their elegant form makes the movement of the avenue run smoothly. Light lines enhance the movement. Bridges decked with bamboo decent to the deep level of the polder water, allowing the visitors to experience this typical Dutch landscape feature which is part of the plan.

The Undulating Bridge establishes a direct connection between the station of Hoofddorp and Beukenhorst South. Coming from the station the bridge enables two possible routes, and therefore has a flared shape. The railing of the bridge gives specific expression to the setting: perforations in the stainless steel walls show a pattern of tree branches. It is the characteristic image of the trees along the Geniedijk silhouetted against the sky polder. LED lighting creates a magical sight in the evening.

勃肯霍斯特南区是哈勒默梅尔围垦地上最新建起的办公园区。它的位置紧靠火车站，又与史基浦机场邻近，这一区域对促进城市的可持续发展有着很好的潜力。园区的主要干线是一条大街：功能上是由于大街位于中心位置，空间上是因为大街中心边缘的双排白蜡树可以作为园区一条绿色的堤坝。沿着大街的方向，流淌着一条宽广的运河，它是这片围垦地景观的又一强有力的结构元素。因为这些长长的线条元素在周围环境中的延续，围垦地在景观中得到很好的强化。

沿着运河可以看到特别精巧的钢板桩，它们标记着那些被改造成城市土地的围垦地区域。钢板桩的一部分被植物覆盖，锈褐色和绿色植物交替的图案，创造出引人注目的效果。这也是对于设计师的可持续性追求的一种表达。

在各种各样的围垦运河的上方，不同的桥梁形成了各有特色的景观元素，并为人们提供了精巧的步行路线。特色鲜明的柯尔顿钢桥横跨运河。在实体上，这些桥梁连接着结实的码头，在细节上，它们优雅的形式使中央大道的动态线条更平滑地延续。光线加强了景观的动感。桥梁的竹制平台正好到达围垦水面的深度，让游客可以体验这种典型的荷兰景观要素是这项设计的一部分。

波浪桥建立了霍夫多尔普火车站和勃肯霍斯特南区之间的直接连接。桥为从车站出来的人们提供了两种可供选择的路线，因此它有一个展开的外形。桥的栏杆对于环境有特别的表达：不锈钢墙上的孔眼形成了树枝的图案。它是沿着格尼迪克大道的树木显示在开垦地最高处的特有图像。LED灯在晚上创造出魔幻的视觉效果。

SQUARE OF KNOWLEDGE
知识广场

Location: Enschede, the Netherlands
Completion: 2012
Design: LODEWIJK BALJON landscape architects
Photography: LODEWIJK BALJON landscape architects

项目地点：荷兰，恩斯赫德
完成时间：2012年
设计师：罗德维克·巴廉景观建筑设计事务所
摄影：罗德维克·巴廉景观建筑设计事务所

The Master Plan for the campus of the University of Twente clusters existing and new buildings, rearranges the traffic structure accordingly and creates a new entry. The basis for the new arrangement is the structure of the landscape and the use of the ecological potential of water.

The Square of Knowledge is the first step in this development; centred on the square are the buildings for Research & Education. The square is focused on the interaction between students and staff. The space is large and open for events. The tent shaped canopy forms a cover that connects the main entrances of the buildings and gives shelter. The object represents the main routing.

The backdrop of trees and hedges combined with scattered benches give ample opportunity for informal meetings in small groups. These benches, red expressive sculptured objects, have a versatile shape that evokes multiple use. Individuals and small groups can use the benches in an active or passive way, ranging from working with a laptop to lounging.

Since the square should be a free open space, a solid green connecting line along one of the edges is designed. The pattern of hedges and the series of narrow tall trees form a green wall and make a connection to the car parks in the surrounding. The watercourse is the boundary of the square on the west side. It is an elongated channel that reflects light and is a vibrant attraction at the edge of the square. The channel is fed by rainwater from the roofs, and that of the square. A filter area has water purifying plants, making also a colourful place along the square. In the second phase, the square will be completed with the extension of the channel. The water is then connected to a deep pond that will be part of the cooling system.

商业与综合景观 · 197

屯特大学校园的总体规划将现有建筑和新建建筑聚合成建筑群，并依此对交通结构重新布置，设计师还为校园添加了一个新的入口。新的规划布局以景观结构和水的生态潜能的运用为基础展开。

知识广场是这一开发项目的第一步，广场位于研究与教学楼群的中心。设计师在对广场进行设计时，注重学生与员工之间的互动。广场的空间大而开阔，适于举办各种活动。帐篷形状的雨棚如同一个盖子，连接了各栋建筑的主入口，并为人们遮阳挡雨。它的行进方向也代表了广场的主要流线。

分散的长椅背靠树木和绿篱，这些元素的结合，为小团体日常的会面提供了足够的空间。广场上有一些如同雕塑的红色长椅，十分富有表现力，它们多面的形状使其具有多种用途。个人或小团体可以坐在椅子上打开笔记本工作，或懒洋洋地躺在上面休息，人们以或积极或消极的方式使用着长椅。

因为广场应该成为一个自由开阔的空间，所以设计师沿着广场一侧的边缘，设计了一条绿色的实体连接线。规则的绿篱和一系列瘦而高

商业与综合景观 • 199

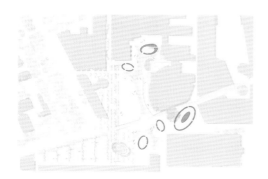

的树木，形成了一座绿墙，并使广场与旁边的停车场建立起了连接。西侧的水渠是广场的边界。它是一条细长的河道，它反射阳光，为广场的边缘增添了活力和吸引力。河道中的水由来自屋顶和广场的雨水补给。设计师还布置了一个有净水装置的过滤区域，它也形成了广场边沿一个多姿多彩的空间。在第二阶段，设计师将延长广场的河道。之后水会被引入到一个深深的池塘，池塘是冷却系统的一部分。

STUDENT RESIDENCE SIEGMUNDS HOF HOUSE 13
齐格蒙德霍夫学生宿舍13号楼

Location: Berlin, Germany
Completion: 2012
Design: die Baupiloten BDA
Photography: NOSHE

项目地点：德国，柏林
完成时间：2012年
设计师：Baupiloten BDA景观事务所
摄影：NOSHE

House 13, "The House for Gardening Friends", has been renovated and rejuvenated, through a sensitive adaptation of the existing fabric coupled with a reinvention of the private and common areas and a new imaginative ecological landscape, in accordance with the desires of the student residents. The new interventions stand out in bold relief against the existing fabric without compromising the building's character.

In order to encourage a mixture of students and a multifaceted living facility, seven different room types have been developed by combining the small existing 10sqm rooms in different constellations. Small single rooms at ground floor have been retained but enriched by either shared pink terraces or direct access to the newly landscaped garden where students can plant and grow their own vegetables. On the upper floors two or three small single units have been combined and installed with a new bathroom and living area which is shared between 1-2 students. The communal kitchens have been expanded and transformed to be the core of the collective, every-day life. Existing transom windows, on the northeast facade, have been supplemented by new large wooden windows which naturally illuminate the kitchen and living space, provide niches for reading and views to the herb garden, outdoor living room and sports ground.

The ecological landscaping and public space around the building form a significant part of the design. New programmatic interventions that play between the qualities of pulsating city and laid back rural life punctuate the area in order to reanimate the previously disused public space. The new town square in front of House13 serves as a focal point for the entire campus, opening itself up towards the city. To the rear, the "outdoor living room" and sports ground provide a quieter recluse from urban life where massive stone seats and terraces can be frequented by those wishing to sunbath or play basketball and oversized external living room lamps create a serene atmosphere for a summer picnic at dawn.

13号宿舍楼又名"园艺之友楼",设计师对其进行了全面的翻修,在原有结构的基础上对私人和公共空间进行了改造,同时还新加入了创造性的生态景观,与住宿学生的爱好相一致。新植入的设施既呈现出明显的特色,又不会影响建筑的品质。

为了鼓励不同的学生入住,宿舍楼设计了7种不同的房间类型,对原有的10平方米房间进行了重新配置。一些一楼的单人房新增了粉色共享露台,另一些则直接与新建的景观花园相连,学生们可自己种植蔬菜。上方楼层配置为两室和三室单位,新增了供学生共用的浴室和生活区。公共厨房进行了扩建,被改造成日常生活的集合中心。东南立面的气窗安装了全新的木窗框,为厨房和起居空间提供了自然照明。学生们可以在这里阅读,也可以欣赏楼下菜园、露天起居室和运动场的景色。

生态景观和公共空间是设计的重要组成部分。设计师将设计介于快节奏城市生活和悠闲的乡村生活之间,使废弃的公共空间重新恢复了活力。13号楼前方新建的城市广场是整个校园的焦点,使校园与城市相连。楼后的"露天起居室"和体育场远离了喧嚣的城市生活。学生可以在大石椅或台阶上享受日光浴,也可以在篮球场挥洒汗水。露天起居室的巨型落地灯为夏日傍晚的野餐营造出宁静的氛围。

商业与综合景观 • 203

KPM-QUARTER – PLOT 5
KPM区第五地块

Location: Berlin, Germany
Completion: 2012
Design: TOPOTEK 1
Area: 2,360sqm

项目地点：德国，柏林
完成时间：2012年
设计师：TOPOTEK 1
面积：2,360平方米

The Spreestadt Charlottenburg, located between the Spree and the Landwehr Canal near the City-West, is a significant 19th Century industrial area. The plans from Ortner and Ortner architects will be developed into a mixed urban area with a high share of residential use. The office of the Federal Health Organisations together with the Royal Porcelain Manufactory (KPM) compose the urban centre of the neighbourhood, thus creating through its free spaces a system of narrow streets and small squares.

The design of the building surroundings relates in its materiality to the existing open spaces of the KPM quarter. The paved areas are made of beige-yellow-small granite stone surfaces, framing harmoniously the covering carpet of granite slabs of the Herbert-Lewin-Platz. Consequently, a generous in-between space opens up and nearby stands a bouquet of black pine, a brass body – which only at a second glance reveals its fountain nature – and a long wooden chaise as furniture. Plantations of oaks and chestnuts surround the whole area. The roof in the 4th floor of the left office building was designed as an accessible terrace surface of concrete paving stone with a central tree element (Euonymus alatus). The direct link to the conference rooms comprises an impressive space designed for festivities with a high representative character, which will be enhanced by the seasonal colour changes of the vegetation and the inspiring views of the surrounding neighbourhood. Out of the direct vision range and protected by the office building for the German Medical Nursery and National Association of Statutory Health, extends its open spaces. The play areas consist of an amorphous form of large soft rubber coating, sand and topographically shaped lawns surfaces that nearly fill the entire outside area. The dark grey coating is the framework of the overall form and at the same time acts as a race and rally track and defines access paths due to its yellow marks. Additionally, it holds the sand play areas in a slightly trough-like form. The lawn area creates a rich contrast between the sand and the rubber coated areas and is shaped as a slight hill. Sunny as well as shady areas allow the use of the open spaces in the various seasons of the year. The designed leisure area offers a variety of means of playing; the markings on the rally track incite to race and run, the sand play area invites to build a castle while the lawn hills tempt lounging.

夏洛滕堡区位于柏林市的施普雷河和兰德维尔运河之间,靠近城西区,是19世纪著名的工业区。Ortner建筑事务所的规划未来会将该区域打造成以住宅为主的混合型城区。德国联邦卫生组织与皇家陶瓷制造公司(KPM)占据了该地区的中心地带,将空间切割成一系列窄街道和小广场。

建筑周边区域的设计与KPM区已有的开放空间相互映衬。铺装区域采用米黄色碎石材质,把中间的花岗石板路夹在中间,从而形成了一个开阔的中间地带,两侧矗立着几棵黑松、一座黄铜喷泉和一条长长的木凳。整个区域被橡树和栗子树所环绕。

办公楼5楼的屋顶被设计成了一个由混凝土铺装的平台,中央设有一个树池(以卫矛为主)。平台与会议室直接相连,共同组成了极富代表性的节庆活动空间。而植物的季节性色彩变化和周边社区的优美景色将进一步提升空间的品质。

德国医疗幼儿园和国际卫生协会办公楼下方是一系列开放空间。游乐区由不定形的软橡胶铺装、沙子和草坪所覆盖。深灰色的橡胶铺装区域是整个空间的框架,同时也通过黄色标记起到了跑道的作用。此外,它还将沙坑游戏区包围起来。草坪与沙子和橡胶形成了丰富的对比,像一座小山。阳光照耀的空间和阴凉空间让开放空间全年都可供人使用。特别设计的休闲区提供了各种各样的活动设施:跑道上的标记鼓励人们赛跑,沙坑游戏区让孩子们堆建城堡,而草坪则吸引着人们休息。

商业与综合景观 • 207

BIOMEDICAL SCIENCE CENTRE
生物医学科学中心

Location: Giessen, Germany
Completion: 2012
Design: Topotek1
Area: 38,000sqm

项目地点：德国，吉森
完成时间：2012年
设计师：Topotek 1景观事务所
面积：38,000平方米

The landscape design concept for the location of the Biomedical Science Centre of Giessen is a main component of the urban design of Seltersberg. A major part of the project covers the modeling of two unequal hills at the entrance and on the central campus area, while its interior is characterised by a distinguished lawn-sculpture, where surrounding paths, spacious stairs, and square like tree elements widen and shape the space.

The composition and generous dimensioning of the landscape elements place them with a central role for both spatial and functional matters in the immediate surroundings as well as in neighbourhood context. The design of the hills and paths follows the curved line's language of the architecture and brings its dynamics to the external space. Articulated lawn-landscape paths define the surrounding areas of the buildings where native multi-stam maple trees are planted in an irregular arrangement. The composition assures and enhances the landscape-quality of space, amplifying the sensation of wideness.

The entrance to the new building of the Biomedical Research Centre is accentuated by a round lawn sculpture at the intersection of Aulweg/Schubertstraße. An equally functional and aesthetically appealing urban furniture enclosures the unity of the project's composition and underlines its useful quality: three custom-made long benches stand along the main paths to the building, framing the central hill. The main square is therefore designed as a pleasurable and unique place of arrival, which together with the gathering space of the cafeteria, stands as the forecourt of the university's engaged daily life.

吉森生物医学科学中心的景观设计规划是塞尔特斯伯格城市规划的重要组成部分。项目的大部分是入口和中心园区的两座景观山丘，内部以独特的草坪雕塑为特色，通过环绕的路径、宽敞的台阶以及广场式树阵来拓宽并塑造空间。

景观元素的组合及其大气的规模使它们处于整个周边环境在空间和功能上的核心地位。山丘和路径的设计延续了建筑的曲线设计，为户外空间带来了动感活力。相互连接的草坪景观路规划出建筑的周边区域，本土枫树在草坪上点缀排列。这种组合保证并提升了空间的景观品质，放大了宽阔的感官效果。

在奥尔维格街和舒伯特斯特拉街交叉口，生物科学研究中心楼的入口处有一座圆形草坪山丘。兼具功能性和美观性的城市景观设施将项目包围起来，突出了它的实用功能：三条定制的长椅坐落在通往建筑的主通道侧面，将中央山丘环绕起来。这样一来，主广场就被设计成了一个宜人而独特的入口空间，与餐厅的聚会空间共同构成了大学日常生活的前院社交场所。

商业与综合景观 • 211

LEONARDO ROYAL HOTEL
莱昂纳多皇家酒店

Location: Munich, Germany
Completion: 2011
Design: Rainer Schmidt Landschaftsarchitekten GmbH
Photography: Raffaella Sirtoli
Area: 7,750sqm

项目地点：德国，慕尼黑
完成时间：2011年
设计师：Rainer Schmidt景观建筑事务所
摄影：拉法埃拉·希尔托里
面积：7,750平方米

The Leonardo Royal Hotel is the first building realised on the site of the former headquarters of the Knorr-Bremse Group. It kicks off the further development of the site in the vicinity of the Munich Olympic Area as a business park. The urban master plan by Rainer Schmidt and JSK Architects provides for a central ribbon of water and pathways, furnished with several spaces for recreation and framed on both sides by buildings. The hotel building is situated on one of the water basins, with a spacious terrace and tiered seating protruding into the pool. The central ribbon runs into a large open space, which is earmarked as a park. The challenge was to create an "anchor spot" with an internal courtyard garden and gateways leading to the future park landscape.

To offset the rigidity of the architecture, the site's open space is covered with expressive "brush strokes". A close-meshed and amorphous net of paths, hidden by tall grasses, shapes the inward and outward garden areas of the horseshoe-shaped building. Looking from the hotel windows, this composition gives the beholder the impression of an abstract painting.

The "brush strokes" demarcate paved areas as paths or sitting zones. The remaining spaces are planted extensively with grasses and flower bulbs in the garden area, gracing it with the bloom of daffodil, giant allium, foxtail lily and reed, depending on the season. As a second layer of planting, various deciduous trees are dispersed over the site in a loose, irregular grid, enhancing the solid effect of the dune-like grass beds. The paths have a light-coloured water-bound surface, widening to small patches for recreation and subdividing the sections planted with reeds. The planted areas are curbed with a granite sett edging. The abstract pattern of paths and patches as well as edgings and plantings create pockets for recreation. On the dark asphalt surface of the hotel's driveway, the "brush stroke" pattern is continued as white forms. The roofs of the hotel also continue the theme with low-growing sedum vegetation and gravel areas also implemented as part of the roof-level design patterns.

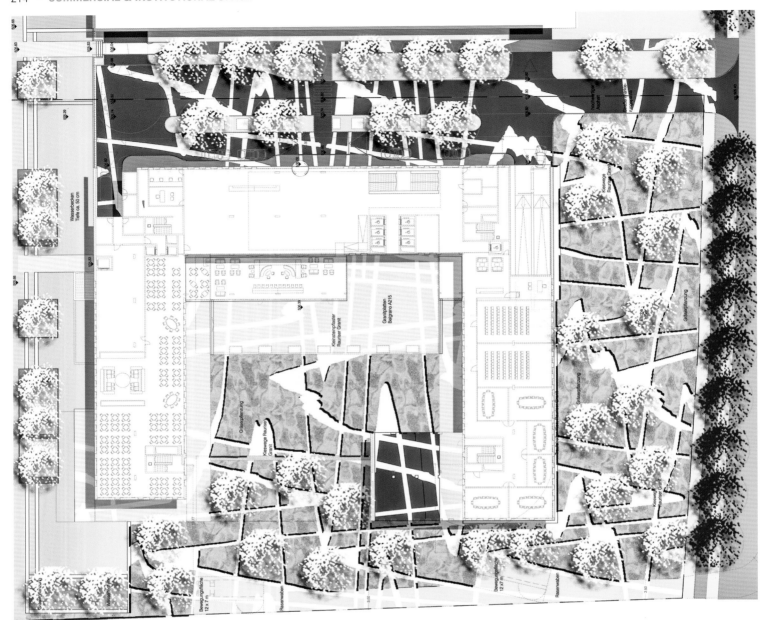

The outdoor area offers peaceful spaces in which to enjoy the changing sunlight, protected from wind – like an oasis in the desert. Its colours, materials and solid planted areas create an atmosphere that complements the functionality of the hotel and defines the hotel as a place to be. It serves as a rearward stage connecting the street, the hotel hall and the garden terrace. It is a place of wellbeing and repose, adding considerably to the value of the site.

The pattern of paths and grasses provides a simple yet powerful and representative image, and furthermore a basic spatial structure that can host different usages and utilities for the various promotions and events the hotel organises. The architecture with its clear proportions and its bright white colour is accented by the pleasant, exciting contrast with the fresh green grass.

It seems that the free forms want to burst out of the orthogonality of the site. By creating recesses and deep spaces and by concentrating and intensifying the impact of plants on the small premises, the means of garden design are used to create a kind of optical illusion. The concept responds to the challenge of an uncultivated site. The contrast between the dense and lush planting and its containment in edged beds, and between the "wild brush strokes" and the delineation of the planted patches with granite setts is a subtle invitation to ponder on the antagonistic relation between nature ("wilderness") and culture ("control"), which can be perceived in this garden in a special way, negotiating with perception continuously.

商业与综合景观 • 215

商业与综合景观 • 217

莱昂纳多皇家酒店是在克诺尔集团总部原址上建成的第一座建筑。它开启了该地块建造商业园区的序幕。由Rainer Schmidt景观建筑事务所和JSK建筑事务所联合打造的城市规划提议建造一条中央水系步行带，其间装点各种休闲空间，两侧是各种建筑。酒店大楼位于水池的中央，配有宽敞的平台和深入水池的阶梯式座椅。设计的主要挑战是打造一个"锚点"，辅以庭院花园和与未来园区景观相连的通道。

未来平衡建筑的棱角，开放空间覆盖着大量富有表现力的"笔触"。不定形道路网络隐藏在高茎草之中，塑造了马蹄形大楼内外的花园区域。从酒店的窗口向外看，整个花园看起来就像一幅抽象画。

"笔触"线条将铺装区域划分为走道和休息区，其他空间则种满了丰富的花草。随着积极的变化，水仙花、巨型葱花、狐尾百合和芦苇次第开放。作为第二层植栽的落叶树分散在各处，形成了松散的不规则网络，提升了沙丘式草坪的实体效果。道路呈浅色水结表面，时而拓宽形成休闲区，时而用芦苇对空间进行划分。植栽区以花岗岩路缘为边界。造型抽象的道路、地块、路缘和植栽共同构成了休闲区。在酒店车道的深色沥青表面上，"笔触"图案以白色线条实现了延续。酒店的屋顶同样通过景天属植物和碎石区域延续了这一主题。

户外休息区为人们提供了享受阳光、躲避风雨的宁静空间，就像沙漠中的绿洲。它的色彩、材质以及植栽区与酒店相辅相成，形成了统一的整体。作为背后的舞台，它与街道、酒店大厅以及花园平台相连。这是一个健康的休憩空间，提升了整个地块的品质。

道路和观赏草所形成的图案简洁而富有表现力，形成了一个基本的空间结构，可以用于酒店的各种宣传活动中。酒店清晰的建筑结构及明亮的白色色调与绿草相互映衬，清新宜人。

自由的造型就像要超脱四四方方的地块之外。错落的空间形式以及鲜明的植物效果采用了花园设计的手法，营造出独特的视觉效果。项目设计将荒芜的地块变成了愉悦的绿洲。茂密的植栽与方正的边界之间的对比、"狂野的笔触"与由花岗岩路缘包围的有序空间的对比都令人不得不深思自然与文明（即野性与控制）之间的关系。花园设计通过独特的方式将两者合二为一。

HOSPITAL LA PAZ
拉巴斯医院

Location: Madrid, Spain
Completion: 2011
Design: Studio A-cero
Photography: Luis H. Segovia

项目地点：西班牙，马德里
完成时间：2011年
设计师：A-cero工作室
摄影：路易斯·H·塞戈维亚

This project has been very special for A-cero, it has been carried out together with the NGO "Juegaterapia". This garden has been designed by A-cero unselfishly. The project is something new for a hospital and it is expected to become very common in hospitals. The aim is to create a playground for the sick children where they can play and stay outdoors. This helps to improve their life quality.

This garden is located in the roof of the Child hospital building of the "La Paz" Hospital. The roof of this Hospital is structured through a set of circular elements that function as different living rooms and play rooms, formally linked to a winding path that runs along the roof, creating an itinerary that welcomes you to get in and to explore all the space.

The roof is carpeted with artificial grass and soft pavement. It is a place thought to meet, read and rest. The porch-tables are very important, circled tables that look like giant mushrooms and provide shadow.

项目对A-cero工作室来说十分特别，它是与非政府组织"Juegaterapia"（游戏疗法）共同合作的。花园的设计由A-cero工作室完成。项目对医院来说显得十分新鲜，但是未来将被广泛推广。项目的目标是为患病儿童打造一个游乐场，让他们可以在户外游戏和停留。这对改善他们的健康十分有益。

花园位于拉巴斯医院儿童住院楼的屋顶。医院的屋顶由一系列圆形元素构成，它们起到了不同的起居室和游戏室的功用。这些元素由一条蜿蜒的小路连接起来，形成了一段吸引人不断探寻的旅程。

屋顶上覆盖着人造草坪和软路面。这是一个聚会、阅读和休息的好去处。门廊桌的设计十分关键，圆桌就像巨大的蘑菇，能为下方带来阴凉。

商业与综合景观 • 221

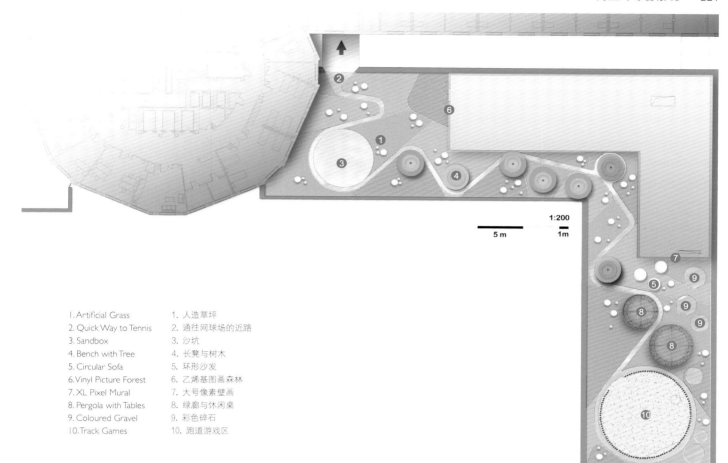

1. Artificial Grass	1. 人造草坪
2. Quick Way to Tennis	2. 通往网球场的近路
3. Sandbox	3. 沙坑
4. Bench with Tree	4. 长凳与树木
5. Circular Sofa	5. 环形沙发
6. Vinyl Picture Forest	6. 乙烯基图画森林
7. XL Pixel Mural	7. 大号像素壁画
8. Pergola with Tables	8. 绿廊与休闲桌
9. Coloured Gravel	9. 彩色碎石
10. Track Games	10. 跑道游戏区

HOSPICE DJURSLAND
日德兰临终关怀医院

Location: Rønde, Denmark
Completion: 2011
Design: C.F. Møller Architects
Area: 1,990sqm

项目地点：丹麦，龙德
完成时间：2011年
设计师：C.F. Møller建筑事务所
面积：1,990平方米

The Hospice Djursland is a palliative treatment facility with room for 15 patients, located in a beautiful landscape setting overlooking the bay of Aarhus. In hospice design, the architect's finest task is to create surroundings which will provide the best possible conditions to promote quality of life, respect and a dignified death. Djursland Hospice is first and foremost a building within a landscape. No matter where you go in the building – the reception area, the garden of the senses, the atriums, the staff room, the lounge, the reflection room or the patient rooms – the beautiful landscape is always present.

The architect have aimed to create a very humane building; by which mean a building which is not an institution, but rather a home which provides adequate physical and mental space for those who will live there in their final time, as well as for their relatives and the staff. The semi-circular layout is to ensure that all patient-rooms enjoy an equally privileged view over the bay, and that they are located within a more private zone in the building, set back from the common rooms. Each room has a private terrace overlooking the landscape, and the section of the roof draws daylight deep into the rooms, providing a skylight over the sleeping area and the bathroom, and a soft curve to the ceiling. The common materials used throughout are copper, oak and glass, which interact beautifully and naturally with the landscape and provide a sense of warmth in the rooms.

The landscape branch of C.F. Møller Architects has designed the landscape and gardens surrounding the hospice, with special emphasis on the sensory aspects of sight, smell, touch and sound, as well as the overall accessibility for patients, even those confined to beds, resulting in a series of soft, rounded shapes and niches joined by green rubber-bitumen surfaces. An extension of the hospice with a new orangery winter garden – filled with exotic plants such as an olive tree, vine, laurel and Japanese bamboo – makes it possible to enjoy the pleasures of outdoor life all year round. Like the rest of the hospice and the sensory garden, the orangery is naturally accessible for both wheelchair users as well as bedridden patients.

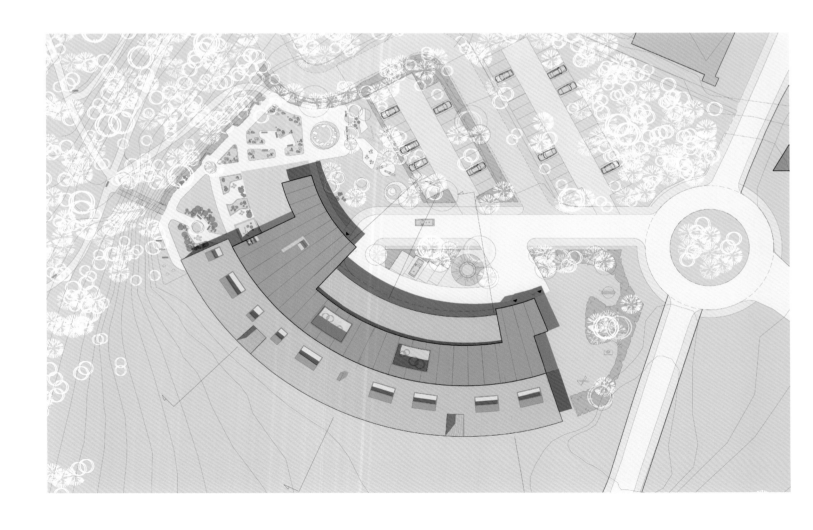

商业与综合景观 • 225

日德兰临终关怀医院是一家采用姑息疗法的治疗机构，可容纳15名患者，四周风景优美，俯瞰着奥尔胡斯湾。在临终关怀医院的设计中，建筑师最重要的任务就是打造尽可能好的环境，帮助患者提高生活品质，使他们度过有尊严的临终生活。日德兰临终关怀医院首先是一座融入风景的建筑。无论是接待区、感官花园、中庭、员工办公室、休息室、反思室，还是病房，优美的景色随处可见。

建筑师的目标是打造一座人性化建筑。这座建筑不是一个机构，而是伴随患者走过生命最后一段路的家园，为他们提供充分的物理治疗和心理治疗。半圆形布局保证了所有病房都能享有海湾的美景，并使它们都坐落在建筑内相对私密的区域，远离公共房间。每间病房都有一个私人露台远眺风景，睡眠区头顶设有天窗，同时还配有浴室和线条柔和的天花板。

项目所使用的主要材料是铜、橡木和玻璃，它们与风景自然优美地相互映衬，为房间营造了温馨之感。

C.F.Møller建筑事务所的景观团队对临终关怀医院的景观和园林进行了设计，特别将重点放在视觉、嗅觉、触觉和听觉的感官体验以及患者进出的方便（甚至考虑到了长期卧床的患者）。最终的景观设计采用柔和浑厚的造型，通过绿色橡胶沥青路面连接起来。

为临终关怀医院的延伸，新的柑橘温室花园内种满了橄榄树、葡萄藤、月桂树、日本竹等异域植物，让患者全年都能享受户外生活的愉悦。与医院的其他部分和感官花园一样，柑橘温室也可以让轮椅和带轮病床自由进出。

AARHUS UNIVERSITY
奥尔胡斯大学

Location: Aarhus, Denmark
Completion: 2013
Design: C. F. Møller Architects
Photography: C. F. Møller Architects
Area: 42,000sqm

项目地点：丹麦，奥尔胡斯
完成时间：2013年
设计师：C. F. Møller建筑事务所
摄影：C. F. Møller建筑事务所
面积：42,000平方米

The campus is situated around a distinct moraine gorge and the buildings for the departments and faculties are placed on the slopes, from the main buildings alongside the ring road to the centre of the city at Nørreport.

The original scheme for the campus park was made by the famous Danish landscape architect C. Th. Sørensen. Until the death of C. Th. Sørensens in 1979 the development of the park areas were conducted in a close cooperation between C. Th. Sørensen, C.F. Møller and the local park authorities. Since 1979 C.F. Møller Architects – in cooperation with the staff at the university – has continued the intentions of the original scheme for the park, and today the park is a beautiful, green area and an immense contribution to both the university and the city in general.

Further to this, landscape works has been undertaken in relation to the continuous integration of new buildings into the existing landscaped moraine valley.

Each landscape project takes its outset, and follows the original intentions with the parks overall plan, the material texture and scope of use. Specific details have been developed in accordance, hereby resulting in new variations and reinterpretations in the use of the park's four main building blocks: meandering yellow brick walls, brick paved surfaces, lawns and oak trees.

The projects have improved access for all user groups to the older buildings, as well as optimising the use of exterior space. Further to the landscape design C.F. Møller Architects has developed bespoke designs for outdoor furniture, signage as well as waste management systems.

商业与综合景观 • 231

校园被一个独特的冰碛石峡谷所环绕，院系和办公楼都建在山坡上，从环形路旁的主楼到诺尔港的城市中心。

这座校园式公园的原始设计由著名的丹麦景观设计师C·索伦森亲自操刀。直至1979年，索伦森去世以前，园区的开发在索伦森事务所、C.F.Møller建筑事务所以及当地公园管理机构的紧密协作下展开。自1979年起，C.F.Møller建筑事务所与学校的教职员工共同延续了园区的原始设计。现在，整个园区环境优美，对学校乃至整个城市都做出了极大的贡献。

此外，景观设计成功的将新建筑与原有的冰碛石峡谷景观融合起来。

每个景观项目都遵循了最初的园区设计方案、材质选择以及视野应用，并相应地添加了一些特殊的细节，在公园内的四个板块形成了新的变奏和解读，即波浪形黄色砖墙、砖铺地面、草坪和橡树。

项目改良了所有用户群进入旧楼的通道，并且优化了室外空间的使用。除了景观设计之外，C.F.Møller建筑事务所还特别定制设计了配套的户外家具、引导标识以及废物处理系统。

NARROW CITY GARDEN
窄城花园

Location: Nijkerk, the Netherlands
Completion: 2012
Design: Buro Harro
Photography: Harro de Jong, Jan Willem Trap
Area: 60 sqm
项目地点：荷兰，奈凯尔克
完成时间：2012年
设计师：布罗·瀚
摄影：瀚·德容，扬·威廉·特拉普
面积：60平方米

This narrow city garden, which borders on the office of the owner on one side, is divided in three parts. The first part, the wooden terrace with the specifically designed bench, is a continuation of the living room. The second part is the flourishing center part of the garden with wild flowers, in front of the office window. The third part is again a terrace and is enclosed by green walls (the fences will completely overgrow in time). The specially designed waterfall is also situated in this part: it supplies a long and narrow water element with water, which is an extra emphasis on the shape of the garden.

The difficulty of this garden, the narrowness, is in this design both solved as well as emphasised in the same time. On the one side the space is closed by a wall that separates it from the neighbours', on the other side by the office-building of the owners, where the space is squeezed a little bit more.

We divided the garden (quit classically) in three comfortable spaces with three characters. The first part functions as an extension of the living room, and looks like that. The wooden floor from inside is continued outside, combined with specifically designed furniture. The bench, with the hedges behind it, divide the long and narrow space in smaller spaces. On the other hand the planks of the terrace (cut in the size of the also very stretched size of the bricks in the walls of this 30's building) emphasize the length and the direction of the space. The second space is a small piece of abundantly blossoming flowers, in front of the office window, which gives a wonderful view from the office inside on the one hand and divides the two terraces in the garden on the other hand. The brick wall on the right is used as a background for two statues by Sjer Jacobs. The third space, the terrace in the back of the garden, is made of bricks, again in direction emphasising the length of the garden and following the lines in the walls. This terrace catches the morning sun.

After dividing the garden in these three comfortable spaces they are reconnected by a waterline, a narrow artificial stream starting in a water element, over the full length of the garden, emphasising the length, narrowness and direction of this

garden again. The waterline (containing even fishes and frogs) adds liveliness to the garden and divides the hard surfaces from the soft surfaces, creating a very narrow border filled with ferns. The vegetation in this small border continues on the walls, giving the small garden a green and pleasant atmosphere even when the biggest parts are terraced.

这座窄城花园的一侧紧挨着主人的办公室，设计师将它分成三个部分。第一部分是一座木质平台，平台从主人的起居室延伸出来，上面摆放着设计别致的长椅。第二部分位于办公室的窗户前，是花园的中心，其中开满繁茂的野花。第三部分又是一座平台，平台被植物绿墙环抱（最终墙体将会完全被植物覆盖）。设计别致的瀑布也位于这一部分：瀑布为一条细长而狭窄的小溪补给了水源，这条小溪再次着重强调了花园细长的形态特征。

狭小是这座公园的设计难点，设计师通过设计，不但解决了这一问题，同时也强调了花园的这一特性。花园的一侧由墙体围合，围墙也将花园与邻居的院子分隔开，花园的另一侧边界由主人的办公建筑界定，那里的空间更加拥挤一点儿。

设计师（脱离古典主义风格）将花园分割成为带有三种不同特性的三个舒适的空间。第一个部分在功能和外形上都作为起居室的延伸。木质的地板从室内延伸到室外，与特别设计的家具结合。长椅以及长椅后面的树篱，将这个细长而狭窄的空间分成更小块的空间。另一方面，平台的木板（木板尺寸被切割成与这座有着30年历史的建筑砖墙尺寸同样的、非常细长的形状）强调了空间的长度和方向。第二个部分是位于办公室窗户前一小块鲜花盛开的空间，一方面，它构成了从办公室内部向外眺望的美妙的视野，另一方面，它将花园中的两个平台分隔开。右侧的砖墙则作为两座吉尔·雅各布斯的雕塑作品的背景。第三个部分是位于花园后边的平台，平台由砖铺成，它再一次在方向上强调了花园的长度，沿续了墙体的线条。在平台上主人可以享受清晨的第一缕阳光。

在将花园分成这三个舒适的空间之后，设计师又把这三个部分通过水流连接，一条狭窄的人造溪流从水景小品处发源，沿着花园长边的方向穿越整个花园，它再一次强调了花园的细长、狭窄和它的方向。这条溪流（其中甚至有鱼和青蛙）增加了花园的活力，并将柔软表面于硬质表面区分开，形成了一个长满蕨类植物且十分狭窄的边界。墙上种植着与狭窄边界上相同的植物，植物赋予这个大部分是平台的小花园绿色而宜人的气氛。

HEIBLOEM
喜花农庄

Location: Brummen, The Netherlands
Completion: 2011
Design: Bureau Poortvliet & Partners garden- & landscape architects bnt
Photography: Mariejanne Vorenholt/Jaap Poortvliet
Area: 75,000sqm

项目地点：荷兰，布鲁门
完成时间：2011年
设计师：比罗·波特福列&合伙人设计事务所
摄影：玛丽珍妮·沃伦霍尔特，亚普·波特福列
面积：75,000平方米

This modest so called "Hall-house-farm" is situated at the convergence of the catchment area of the river IJssel and the edge of the Veluwe, at the boundary between clay and sand in the area of Oeken (municipality of Brummen). The area of the plot is 7.5 hectare. The farm is mentioned on a fragment of a map dated 1830. The name "Heidebloem" or "Heatherflower" is used in a topographical map from about 1900. The owner has lived in the farmhouse for many years, but decided that a radical renovation and reconstruction was needed. The country character of the farmhouse had been seriously impaired by a modernisation carried out by previous inhabtants.

The architect first suggested a new rebuild but the clients thought this out of character with the country surroundings. At this stage designer were brought into help think about the integration of the area into the countryside and the development of a landscape plan.

The surroundings, a lovely area with intermingling wooded areas, hedges, hawthorn hedges and fields are its strongest point. The farmhouse lies nestles in the landscape, with, on one side an old orchard and on the other side an area with groups of ancient oaks and sweet chestnuts each with an average trunk circumference of one metre. The farm complex was fairly typical, consisting of a farmhouse, a number of barns in use as a garage/storage, stable/hayloft, chicken run, and bakery. Subsequently it was decided to remove the bakery as it interfered with the desired view. On the street side of the farmhouse the typical barn doors used by the inhabitants, both human and animal were missing. The front door was not inviting and difficult to get to. The entrances to the garage and the front door were hidden from the road and facing the open country.

The characteristic features of this type of farmhouse in this area of the countryside were again restored during the renovation. The barn doors are now used as the entrance to the farmhouse and the area in front is used for cars. The entrance to the garage and the arrangement of the yard is in keeping with this arrangement. The

farmhouse and the garage are joined together by the garden wall which creates a clear division between "front" and "back".

To the rear the farmhouse has been enhanced by the formal garden, lawn, orchard, vegetable garden, stable and chicken run. Young trees have been planted within the old orchard. The free standing groups of old trees have been integrated into the garden design.

A very old and characteristic cherry tree is now an imposing focus point between two long flower borders. The lawn at the end merges seamlessly into a flower meadow which continues up to the boundary with the wood.

这座端庄的"庄园-农舍-农场"综合体,位于费吕沃边境、艾塞尔河下游汇流区,在(布鲁门市)欧肯区的黏土和沙地的交界处。这一区域面积7.5公顷。有关农场的记载最早出现在一张19世纪30年代破碎的地图上。它的名字"喜花"或"石楠花"在1900年前后的地形图上有标注。农场主在农舍里居住多年,但是现在他认为有必要对农场进行一次彻底的翻修和重建。由于过去居民实行的现代化建设,农舍的乡村特征遭到了严重的破坏。

最初,建筑师建议对房屋进行重建,但是客户认为新建的房屋与周边的乡村环境格格不入。眼下客户找到设计师,希望设计师帮助他们对这里的景观进行规划,使这一区域与周围乡村环境得到整合。

农舍周围的环境非常优美,这一区域混合生长着多种植物,其中森林区、树篱、山楂树篱和农田是最重要的景观节点。农舍依偎在景观中,如同一个巢穴,农舍的一侧是一个古老果园,另一侧则生长着成群的老橡树和欧洲栗,平均每棵树的胸径为一米。这个农庄是一个相当典型的综合体,它由一座农舍和几座作为车库或储藏室的谷仓,以及马厩、干草棚、养鸡棚和面包房组成。随后设计师决定移除面包房,因为它阻碍了主人期望获得的景观视线。在农舍沿街的一侧,居民使用典型的谷仓门,现在这里不再有人养殖牲口。前门并不明显且难以到达。在公路上看不到车库的入口和前门,因为它们朝向旷野。

经过翻修,这里典型的乡村农家特征再一次被还原。谷仓门现在作为农舍的入口,门前的区域用来停车。车库的入口和院子的布置与这种布局保持协调。农舍和车库通过花园围墙连接在一起,在"前"、"后"之间形成了一个清晰的界限。

农舍的后方则通过规则式的花园以及草地、果园、菜园、马厩、养鸡棚得到加强。老果园中新栽植了幼树。自由生长的老树群组与花园的设计融为一体。

现在,在两条长长的花境之间,一棵非常年长的樱桃树成为特有的景观节点。草地末端与花草地无缝地融合,花草地继续延伸向农庄边界的树林。

WIJSSELSE ENK
威斯克安科农庄

Location: Apeldoorn, the Netherlands
Completion: 2011
Design: Bureau Poortvliet & Partners garden- & landscape architects bnt
Photography: Mariejanne Vorenholt/Jaap Poortvliet
Area: 63.400sqm

项目地点：荷兰，阿陪尔顿
完成时间：2011年
设计师：比罗·波特福列&合伙人设计事务所
摄影：玛丽珍妮·沃伦霍尔特，亚普·波特福列
面积：63,400平方米

This historic, so called "Hall-house-farm" from 1838, a local monument, lies in the part of Wenum Wiesel (in the municipality of Apeldoorn) that is called Huisakkers. The site is situated on the border between de Wieselsche Enk and the sandy valley of the Wenum's stream, which forms the border of the property on the south side of the farmhouse. A public road runs through part of the private grounds. The area is 63,400 sqm. Buildings, wooded areas and fields combine to create a particularly attractive landscape, but due to a lack of maintenance in the past, the farmhouse and barns have been completely renovated and restored both inside and outside, with the construction of a haystack as the finishing touch. This restoring both spatially and functionally the inherited estate.

The designers were asked to prepare a complete vision and landscape plan. Tackling the lack of maintenance of the linear landscape elements had a high priority. A detailed estate and garden plan was prepared as part of the landscape plan.

The cultural history of the area has been charted out using a detailed map from 1832, topographical maps from about 1850/1875 en subsequent years and aerial photos from 1938 to the present day. The conclusion was that moves to larger scale working and land usage had deleteriously impacted on the spatial qualities of the area. A renovation plan was prepared using this data. In addition, two new landscaping elements were added: a walnut avenue, a group of linden trees and a fruit orchard. A number of sight lines from the farmhouse show the historical stratification: landscape – cultural history – garden.

The existing farmhouse has been cleaned up with two valuable elements now dominating the estate; the rising linden at the front and the fully grown beech hedge on the boundary. These form the boundaries of the 60 x55 metre living area. The use of the farmhouse does not conform to the traditional arrangement of a farmyard into a "frontside" and a "backside". At the front there is the bakery, which is now used as an office by the wife of the owner, while part of the front of the house is used as an office by the husband. However, the organisation of the outside

space reflects the traditional farmyard very well. At the front of the house there is a flower garden and a large lawn which seamlessly flows into flowery meadows. Sheltered between the farmhouse and the barns one finds the patios, the flower picking garden, the herb garden and a contemporary well. The separation between the domestic and working areas is characteristic and is further accentuated by the height differences between the barns and the farmhouse. The height difference of 60cm within the farmhouse is repeated outside and supported by cortensteel walls.

这个建于1838年的"庄园-农舍-农场"综合体，是当地著名的历史遗迹，它位于（阿陪尔顿市）维姆维塞尔的胡萨克斯。基地位于威斯克安科和维姆的溪谷沙滩之间，维姆溪流划定了农舍土地南侧的边界。一条公路穿越这块私人土地的一部分。这块基地的面积为45,000平方米。这里的建筑、森林区和农田结合在一起，形成了一幅特别引人入胜的景观，但是由于过去疏于维护，一些部分受到破坏，因此主人对农舍和谷仓进行了彻底的翻修，翻修后，建筑的内部和外部都得以恢复，干草堆的建设是修复工程的最后一个步骤。至此，这块传承历史的土地，在空间和功能上都得到了复原。

主人要求设计师制定一个完整的视觉和景观规划方案。处理疏于维护的线性景观元素作为设计重点，被设计师提到了最高优先级。同时，作为本次景观规划的一部分，设计师也为房屋和花园制定了详细的设计方案。

这一区域的文化历史可以追溯到1832年，那时的一张详细地图就对这一地区有标注，大约从1850年或1875年起，出现了对这里的地形详细描绘的地形图，从1938年至今的航拍照片也显示了从此之后这一区域的布局。设计师的结论是，大尺度的规划和土地使用方式，会对这一区域的空间品质产生破坏性影响。他们在设计修复方案时参考了这些资料。此外，设计师在基地中增加了两个新的景观元素：一条胡桃林大道、一组菩提树和一个水果园。从农舍出发，延伸出的若干条景观视线，展示出了这里的历史层次：景观——文化历史——花园。施工人员对原有的农舍进行了翻修，现在有两个重要的元素起到了景观主导的作用：农舍前方生长的菩提树和边界茂盛生长的山毛榉树篱。树篱形成了这块60乘55平方米生活区域的边界。

农舍的使用与传统农家庭院分为"正面"和"后面"的布局不同。在农舍的前方有一个面包店，现在它是农舍主人妻子的办公室，农舍前面的部分是丈夫的办公室。但是，外部空间的组织非常好的继承了传统农家庭院的特征。农舍的前面有一个花园和一块大草坪，草坪与花草地无缝地衔接。隐藏在农舍和谷仓之间的，是露台、采摘花园、草药园和一口现代井。生活和工作区域之间有明显的区隔，这种分隔通过谷仓和农舍之间的高差进一步得到加强。农舍室内标高有60厘米，外部平台也是相同的高度，平台由柯尔顿钢墙支撑。

私人花园与住宅景观 • 247

GARDEN OF A VILLA IN BERLIN-DAHLEM
柏林-达勒姆别墅花园

Location: Berlin, Germany
Completion: 2012
Design: glasser und dagenbach landscape architects
Photography: Udo Dagenbach
Area: 1,500sqm

项目地点：德国，柏林
完成时间：2012年
设计师：glasser und dagenbach景观事务所
摄影：乌多·达根巴赫
面积：1,500平方米

The basic idea of the design principle for the home garden was shield to the surrounding neighbouring areas and structures from the garden and the structuring of the garden itself in diverse areas, which respectively enables the use for all family members.

The crucial idea of the reorganisation of the garden was implemented using linear and selective elements from plants (Fagus sylvatica, Carpinus betulus, Buxus sempervirens) and stone (limestone walls, limestone cuboids, gabion walls of limestone, terrace slabs of limestone, basalt stone paving). The repetition of the stone elements and the play of colours in the plants composed from a mixture of red-leaved trees as solitaire and hedge lines, reinforces the impression of the garden by the appearance linear tree plantings of high hornbeam strains and hornbeam walls along the property border. In general, only white and blue purple flowering shrubs, perennials and bulb plants were used, thus stimulating a greater depth effect of the landscape, through the selective use of adequate flower colours. The interaction of the recurring materials and plant structures in different dimensions results in a spatial expanse. The staging of the different areas in the garden under a parent design theme was the main objective in the design idea.

这个家庭花园的基本设计理念是使其与周边的区域隔开并在花园的不同区域设计出各种功能空间，供所有家庭成员使用。

花园重新布局的关键在于精选的植物（欧洲山毛榉、欧洲鹅耳枥、锦熟黄杨）和石材（石灰岩墙、石灰岩块、石灰岩石笼墙、石灰岩平台板、玄武岩铺面）等直线形元素。石材元素的反复出现以及植物的丰富的色彩（综合了多种红叶树木）通过高大的角树和墙壁的线条突出了花园给人的印象。花园设计在整体上仅采用了白色和蓝紫色的花丛、多年生植物以及鳞基植物。精心挑选的花卉色彩组合在一起，形成了一种深邃的景观效果。同样的材料和植物结构以不同的比例反复出现，它们之间的互动形成了一种空间扩张。花园中不同区域在统一主题下的分阶段性呈现是整体设计的主要目标。

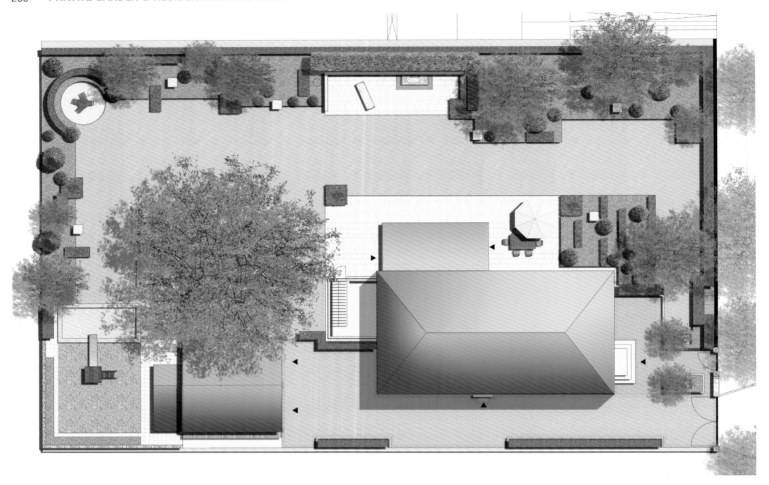

私人花园与住宅景观 • 253

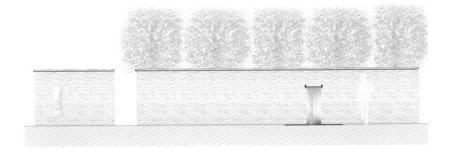

1001 NIGHT HOUSE
一千零一夜住宅

Location: Madrid, Spain
Completion: 2012
Design: Studio A-Cero
Photography: Luis H. Segovia
Area: 7,000sqm

项目地点：西班牙，马德里
完成时间：2012年
设计师：A-Cero工作室
摄影：路易斯·H·塞戈维亚
面积：7,000平方米

It is a single family house located in a development in the outskirts of Madrid. Over a plot of 7,000m^2 the building, with 2,100m^2, rises with a high standard design answering the owners' requirements.

The access, through the development walkway, is placed on a higher level. Here is the property, partially hidden by many curved walls that seem to elevate from some water sheets over a stone covering in white, grey and black shades, placed on purpose as a part of the landscape in this area of the plot. Besides its sculptural features, typical of the A-cero style, this side of the façade expects the integration of the building in the surrounding environment. A wide stone path, with water sheets on both sides, lead us to a huge black glass door that gives us access inside the property. In the garden, following the wishes of the owners, there are palms, pome granate trees and Middle East vegetation.

The rear facade of the house, the most visible, makes the most of the slight slope of the plot, where there is the porch, a pool and the garden. Almost all the views from the different rooms of the property are focused here, as the views of the lakes in the common areas of the development.

All the building is dressed in "black villar granite stone". In this part of the property big windows, with hidden woodwork, are opened, achieving a lot of light for the inside space. In the porch, the window in the main living room, of 10 metres, is automatically hidden, connecting indoors and outdoors.

The passable area is made of white marble, the vase in the pool of blue gressite. The outside furniture is from the Rest collection by A-cero In.

这是一座位于马德里城郊住宅区的独立式家庭住宅。项目总占地面积7,000平方米，住宅建筑面积2,100平方米，根据业主的需求打造了高端大气的设计。

住宅的入口设在相对较高的位置，与住宅区的人行道相连。住宅半隐藏在多面弧形围墙之后，这些围墙好似从水中卷曲而上，水下的地面铺设着白色、灰色和黑色的碎石，形成了独特的景观效果。除了A-Cero工作室的经典雕塑性特征之外，前院的设计还让建筑巧妙地融入了周边环境。宽阔的石板路在两侧水池的伴随下引导着我们走进黑色玻璃大门，进入宅邸内部。根据业主的要求，花园里种植着棕榈树、大果石榴树和中东植物。

住宅的后院充分利用了场地的缓坡地势，将门廊、游泳池和花园结合起来。几乎所有的房间都享有后院及住宅区公共区域湖泊的景色。建筑整体由黑色比利亚尔花岗岩覆盖。宽大的窗口将木框隐藏起来，为室内空间带来了充足的采光。门廊处主客厅的窗户长10米，自然而然地将室内外空间连接起来。室外通道采用白色大理石铺装，而泳池底部则铺设着蓝色瓷砖。户外家具来自A-cero In的Rest休闲系列。

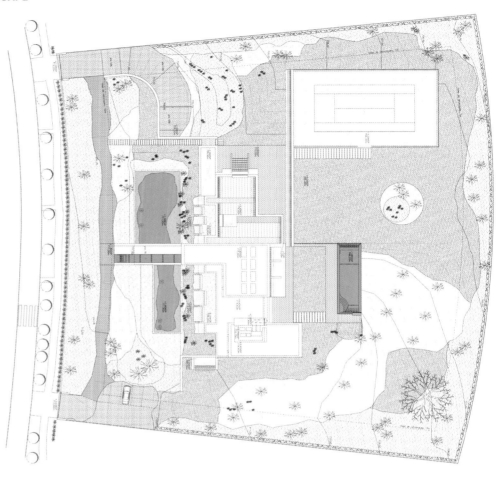

HOUSE IN SOMAGUAS
索玛古亚斯住宅

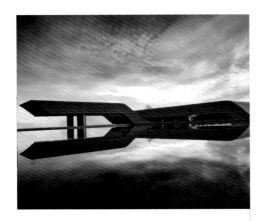

Location: Madrid, Spain
Completion: 2011
Design: Studio A-cero
Photography: Joaquin Torres

项目地点：西班牙，马德里
完成时间：2011年
设计师：A-cero工作室
摄影：杰奎因·托雷斯

The Spanish architecture firm A-cero, directed by Joaquin Torres, has built a new house in the Madrid outskirts that synthesises the evolution of the studio's signature design language and its technical experimentation over the last years. The house can be aesthetically inscribed in the series of projects made by the studio since its international expansion, in places like the Dominican Republic and Dubai, presenting a greater spatial complexity and use of shapes that underlines the relation between A-cero's architecture and contemporary sculpture.

At first impression the house clearly shows its intentions, with the dominance of stylised curves and bold shapes that relate harmonically to its natural context while keeping a clearly modern character. The horizontal shapes pile up one on another, creating a stratified building that seems to emerge from the earth like a natural formation, the façades are treated with a texturised dark concrete, completing the mineral analogy. In this capacity of being at once natural in its matter and artificial in its forms, the house reminds of the work of minimalist sculptors like David Nash, or a piece of land art.

The interior contains a varied programme, solved with a very complex array of spaces with different heights and levels, as well as the particular shape of some of the rooms. The lower level contains the main hall – covered by a curved ceiling that accentuates its relevance – living and dining rooms, master bedroom, gym, interior pool, kitchen and service areas. On the upper level is located a painting studio, under a long curved ceiling, flooded with natural light and the best views over the surrounding landscape. The basement is dedicated to health and leisure, with a bar, games room, chill out, massage room, projection room, cellar and gym.

The spaces are freed of columns and other elements that would alter its fluidity and openness, light materials have been used in the interior design to improve this aspect. The floors are covered with big format white ceramic tiles and the bathrooms are finished in white aluminum.

私人花园与住宅景观 • 259

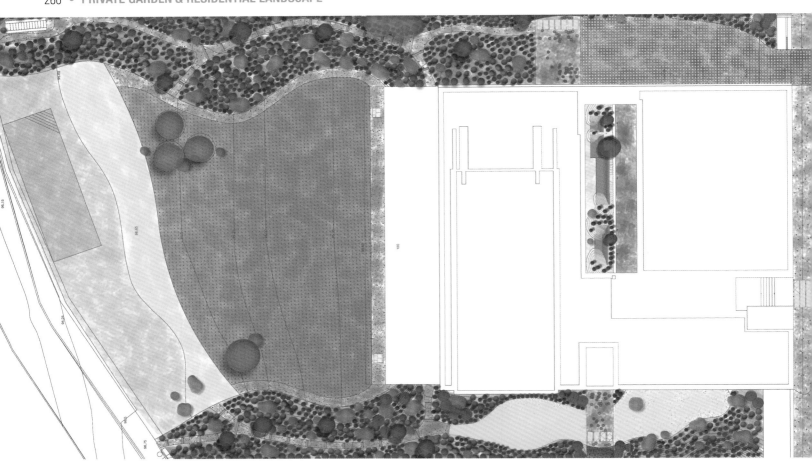

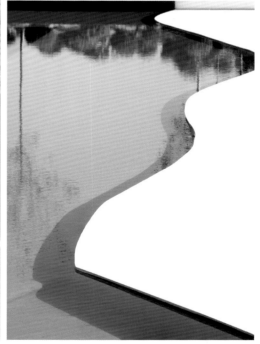

西班牙建筑事务所A-cero在西班牙马德里郊区设计了一座新住宅。住宅综合了A-cero工作室多年以来形成的标志性设计和技术试验成果。住宅体现了工作室进军国际（多米尼亚、迪拜等）市场以来的设计美学，呈现了更复杂的空间组合与造型应用，突出了A-cero工作室的建筑与现代雕塑之间的关系。

住宅很明显地彰显着自己的风格：独特的曲线与大胆的造型同自然环境和谐地联系起来，同时又保持了鲜明的现代特色。水平造型相互堆叠，形成了富有层次感的建筑，好似自然形态中生成的地块一样，建筑外墙采用深色纹理混凝土进行处理，更具矿物感。建筑本质自然，造型人工，令人想起极简主义雕塑家戴维·纳什的雕塑作品，或者是一件地景艺术品。

室内设计包含各种各样的设计规划，空间序列的高度与层次各有不同，形成了各种不同形状的房间。下层空间包含主厅（覆盖在弧形天花板下）、客厅、餐厅、主卧、健身房、室内游泳池、厨房和服务区。上层空间是一间画室，享受着自然采光和周围优美的风景。地下室专用作健康休闲，设有吧台、游戏室、冷库、按摩室、放映室、酒窖和健身房。

整个空间采用无柱结构，尽显流畅感与开放感。室内设计全部选择了轻质材料。地面覆盖着大块白色瓷砖，浴室则采用白色铝材装饰。

THE GARDEN OF WATER HOUSE
水屋花园

Location: Altea, Spain
Completion: 2013
Design: Esculpir el Aire
Photography: José Ángel Ruiz Cáceres
Area: 1,200sqm
项目地点：西班牙，阿尔特亚
完成时间：2013年
设计师：Esculpir el Aire景观事务所
摄影：Esculpir el Aire景观事务所
面积：1,200平方米

The landscaping project and the architectural project operate together to obtain a good adjustment to the existing topography in El Paradiso, Altea, Alicante, Spain.

The architectural design defines two ways of movement: an outdoor stairs between the house and the main street, and an outdoor ramp of vehicles at the back front.

The landscape design of "The Garden" works, from the architectural design, with three main ideas:
1. To dilute subtly the limits between the existing vegetation mountain and the new areas of subtropical plants proposed. The subtropical species give way to mountain plants and vice versa. The Garden works as a continuous and rhythmic wide space. The entrance takes rhythm and colour through several palm trees and shrubs (Phoenix roebelenii, Bouganvillea spp. Hedera helix, Hibiscus spp, etc.). Large specimens of Cyca circinalis and Cyca revoluta work as milestones in both the entrance and the transition zone between the slope and parking access. In the north also it works with subtropical plants, in this case, Schefflera actinophylla, Howea forsteriana, Nephrolepis cordata, etc. for adapting the white walls and allowing the use of green in a dark area.
2. To introduce the mount inside the plot and dilute the boundary between landscape and housing. In this case, the use of Pennisetum rubrum, Gaura lindheimeri and Miscantthus spp. introduces the autochthonous vegetation to get the integration.
3. To provide privacy to the house without losing the spectacular views that it offers. To get this end, it is planted a forest of Ficus nitida next to the pool, which is conveniently pruned, covering the bottom of the valley, closing the outdoor seating area from prying eyes, but keeping the beautiful views of the surroundings.

私人花园与住宅景观 • 263

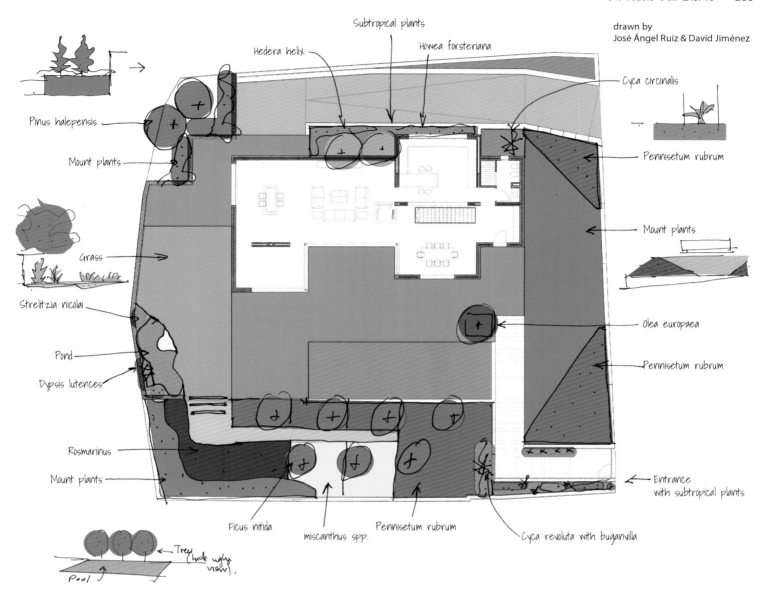

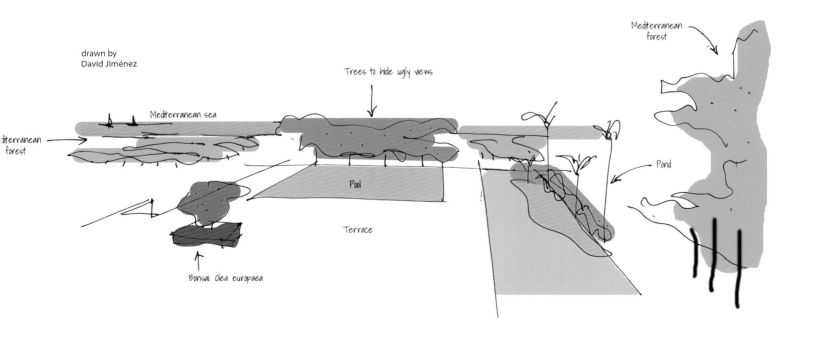

项目的景观设计与建筑设计是同步协调进行的,以实现项目与其所在地西班牙阿尔特亚帕拉迪索地区的地形相适应。

建筑设计以两种移动方式为主:住宅与主街道之间的户外阶梯以及后院的户外车辆坡道。

花园的景观设计拥有三个主要设计概念:

1. 缓解原有的山地植物与新添的亚热带植物之间的突兀感,使二者相互融合,和谐统一。花园是一个连续且富有节奏感的开阔空间。入口处通过一些棕榈树和灌木(江边刺葵、簕杜鹃、常春藤、木槿等)来营造层次感和色彩感。叉叶苏铁和琉球苏铁等大型物种被用作入口和坡道与停车通道之间过渡带的标志。住宅北面同样以亚热带植物为主,包括鹅掌柴、金帝葵、肾蕨等。它们与白墙相互映衬,实现了绿色在暗区中的运用。

2. 在住宅场地中引入山丘,缓解景观与住宅之间的界限。狼尾草、山桃草和芒属植物等本土植物都促进了这种融合。

3. 既保证住宅的私密性又不使其丧失自身享有的景观视野。设计师在泳池边栽种了一片榕箭竹。它们便于修剪,覆盖了山谷的底部,将户外休息区掩藏起来,同时又保持了周边的良好视野。

CASA DE CAMPO
田园之家

Location: Santiago, Spain
Completion: 2011
Design: Rencoret y Rüttimann Arquitectura y Paisaje
Photography: Cecilia Rencoret Ríos
Area: 50,000sqm

项目地点：西班牙，圣地亚哥
完成时间：2011年
设计师：Rencoret y Rüttimann建筑与城市规划事务所
摄影：塞西莉亚·伦克里特·里奥斯
面积：50,000平方米

The landscape consists of a large central valley that runs from North to South, surrounded by two mountain ranges: the Andes Mountains to the East and West coastal mountain range. The Green Valley, contrasts with the slopes of the hills that are more dry and abrupt. Within this horizontal landscape stand out the ancient country houses that are made of adobe, with roofs of clay tiles, corridors and courtyards. Around them there are often planted exotic trees parks, where palm trees, cedars, pine, Oaks and other species whose silhouettes stand out against the sky. The geometry of the agricultural landscape is of straight lines, because its shape is determined by the logic of irrigation channels and farming systems. They are large panels of varied colours and textures intercalated by roads, canals and groves. The commission consented to restore a traditional house located in this context, along with its park and garden, who had spent almost 40 years ago abandoned. The challenge was to integrate elements of the identity and tradition to the local historical and aesthetic canons current design.

The house is situated on a plot of land of 5 ha, surrounded by crops. To the south is the access and the planting of trees that would give rise to the park (40,400 m2). Toward the north, where the windows are opened and corridors of the house, would be the garden, in an area on a regular basis fenced (7,600 m2). The construction was done in stages, first the restoration of the house, then the garden and the courtyard and finally the park that is still under construction.

The garden project proposes the reinterpretation of the traditional Chilean model and integration into the existing landscape. Irrigation techniques and traditional species were used, but they were planted in non-traditional form: large panels of colour of regular shapes, making an analogy to the cultivated fields. The shape of the garden is based on the geometry of the House, crops and flood irrigation system. Orthogonal lines which determined areas of plantation, circulation and irrigation channels have been traced. The landscape designers created a horizontal plane, punctuated by some trees, which continues towards the agricultural plantations which surround the garden and extend across the Valley topping in the cordillera.

The feeling of being in the Valley surrounded by mountains in this piece of planted landscape is perpetuated.

The park develops in an area of existing trees that it was necessary to recover. It considered to complete the plantations with trees and flowers and lodge several programmatic elements as court of tennis, court of soccer, areas of being, lookouts and a giant chess. For this a new footpaths structure is established on the place with proper laws and geometry that support and legibility give him to the program. This structure of straight lines, like the channels and ways, is born from the accesses to the house and intercepts the trees masses happening between lights and shades.

巨大的中央山谷由南向北延伸，被两座山脉所环绕：东面是安第斯山脉，西面是沿海山脉。绿谷与略显干枯陡峭的山坡形成了鲜明的对比。在这片风景线中，一群有土砖建成的古老的乡村住宅脱颖而出，它们拥有陶瓦屋顶、走廊和庭院。住宅四周通常环绕着具有异国风情的花园，种植着棕榈树、香柏、松树和其他物种，在天空中勾勒出剪影。农田风景是直线条的，由灌溉渠和农耕系统的逻辑线条所决定。它们像五颜六色的画板，与公路、运河和树林交织起来。项目的任务是修复一座位于这种环境下的传统住宅及其配套的园林，后者已经荒废了近40年。项目所面临的挑战是将住宅自身的标志性元素和当地的历史传统设计融合起来。

住宅所处的土地面积约5公顷，四周环绕着农田。入口设在南面，园林面积40,400平方米，种植着各色树木。北面的窗户和走廊正对7,600平方米的花园，外面由围墙环绕。项目施工分期完成，首先是房屋的修复，然后是花园和庭院的改造，最后是仍在建设中的公园。

花园项目重新诠释了传统的智利花园模式，使其融入了原有的景观。景观设计师选择了以非传统形式来实施了灌溉技术并种植了传统物种：大面积的规则造型的色块与四周的农田遥相呼应。花园的造型以房屋、农田和漫灌系统的结构为基础。垂直线条决定了植栽、交通和灌溉渠的设计。景观设计师打造了一个水平面，由树木点缀，一直延伸到环绕花园、穿越山谷延伸到山顶的农田。设计突出了置身于众山环绕的山谷之中的感觉。

公园建在一片亟待恢复的林地之上，将由树木、花草和一些如网球场、足球场、空地、棋盘公园等的功能元素组合而成。公园中还新修了一条小路，将各个功能区连接起来。沟渠、道路等直线景观元素全部始于住宅的入口通道，在光影中与花草树木相互交错。

Index 索引

OKRA landschapsarchitecten bv
www.okra.nl
mail@okra.nl

Carve
www.carve.nl
info@carve.nl

Büro Grün plan
www.gruen-plan.de
info@gruen-plan.de

A24 Landschaft
http://a24-landschaft.com
info@a24-landschaft.de

Rehwaldt Landschaftsarchitekten
http://rehwaldt.net
mail@rehwaldt.de

TOPOTEK1
http://topotek1.de
topotek1@topotek1.de

Glaßer und dagenbach landscape architects bdla
www.glada-berlin.de
info@glada-berlin.de

gruppe F Landschaftsarchitekten
http://gruppef.com
info@gruppef.com

POLA
www.pola-berlin.de
mail@pola-berlin.de

bbzl
www.bbzl.de
application@bbzl.de

Hager Partner AG
www.hager-ag.ch
info@hager-ag.ch

Balmori Associates
www.balmori.com
info@balmori.com

EFFEKT
www.effekt.dk
hello@effekt.dk

CEBRA
http://cebraarchitecture.dk
cebra@cebraarchitecture.dk

DE URBANISTEN
www.urbanisten.nl
info@urbanisten.nl

Posad
http://www.posad.nl
mail@posad.nl

RO&AD Architecten

www.ro-ad.org

informatie@ro-ad.org

Die LandschaftsArchitekten

www.die-landschaftsarchitekten.ch

info@dielandschaftsarchitekten.de

Rainer Schmidt Landschaftsarchitekten GmbH

www.schmidt-landschaftsarchitekten.de

mail@rainerschmidt.com

JUNQUERA arquitectos

http://junqueraarquitectos.com

estudio@junqueraarquitectos.com

Esculpir el Aire

http://esculpirelaire.com

esculpirelaire@ctaa.net

Benedetta Tagliabue

http://www.mirallestagliabue.com

info@mirallestagliabue.com

Thorbjörn Andersson

http://thorbjorn-andersson.com

thorbjorn.andersson@sweco.se

Buro Lubbers landscape architecture and urbanism

http://www.burolubbers.nl

info@burolubbers.nl

Karres en Brands landscape architecture + urban planning

http://www.karresenbrands.nl

info@karresenbrands.nl

LODEWIJK BALJON landscape architects

www.baljon.nl

landscape@baljon.nl

die Baupiloten BDA

http://www.baupiloten.com

post@baupiloten.com

Studio A-Cero

www.a-cero.com

a-cero@a-cero.com

C.F. Møller Architects

http://www.cfmoller.com

cfmoller@cfmoller.com

Buro Harro

www.buroharro.nl

mail@buroharro.nl

Bureau Poortvliet & Partners

http://www.poortvlietenpartners.nl

info@poortvlietenpartners.nl

Rencoret y Rüttimann Arquitectura y Paisaje

http://www.rencoretyruttimann.cl

proyectos@ryrarquitectos.cl

图书在版编目（ＣＩＰ）数据

现代欧洲景观设计 /(德) 达根巴赫编；常文心译.-- 沈阳：辽宁科学技术出版社, 2015.9
ISBN 978-7-5381-9250-6

Ⅰ. ①现… Ⅱ. ①达… ②常… Ⅲ. ①景观设计－作品集－德国－现代 Ⅳ. ①TU-881.516

中国版本图书馆 CIP 数据核字(2015)第 108094 号

出版发行：辽宁科学技术出版社
　　　　　（地址：沈阳市和平区十一纬路29号　邮编：110003）
印　刷　者：利丰雅高印刷（深圳）有限公司
经　销　者：各地新华书店
幅面尺寸：225mm×285mm
印　　张：17
插　　页：4
字　　数：100 千字
出版时间：2015 年 9 月第 1 版
印刷时间：2015 年 9 月第 1 次印刷
责任编辑：马竹音
封面设计：周　洁
封面设计：周　洁
责任校对：周　文

书号：ISBN 978-7-5381-9250-6
定价：298.00 元

联系电话：024-23284360
邮购热线：024-23284502
http://www.lnkj.com.cn